Wireless Networks

Series Editor

Xuemin Sherman Shen, University of Waterloo, Waterloo, Canada

The purpose of Springer's Wireless Networks book series is to establish the state of the art and set the course for future research and development in wireless communication networks. The scope of this series includes not only all aspects of wireless networks (including cellular networks, WiFi, sensor networks, and vehicular networks), but related areas such as cloud computing and big data. The series serves as a central source of references for wireless networks research and development. It aims to publish thorough and cohesive overviews on specific topics in wireless networks, as well as works that are larger in scope than survey articles and that contain more detailed background information. The series also provides coverage of advanced and timely topics worthy of monographs, contributed volumes, textbooks and handbooks.

Changyan Yi · Ruoyang Chen · Jiayuan Chen ·
Xiaolong Li · Jun Cai

Self-Evolving Digital Twin Over Wireless Networks

Dynamic Twin Construction and Service Interaction

Springer

Changyan Yi
College of Computer Science
and Technology
Nanjing University of Aeronautics
and Astronautics
Nanjing, Jiangsu, China

Ruoyang Chen
College of Computer Science
and Technology
Nanjing University of Aeronautics
and Astronautics
Nanjing, Jiangsu, China

Jiayuan Chen
College of Computer Science
and Technology
Nanjing University of Aeronautics
and Astronautics
Nanjing, Jiangsu, China

Xiaolong Li
School of Computer Science
Hunan University of Technology
and Business
Changsha, Hunan, China

Jun Cai
Department of Electrical and Computer
Engineering
Concordia University
Montreal, QC, Canada

ISSN 2366-1186 ISSN 2366-1445 (electronic)
Wireless Networks
ISBN 978-3-032-08119-3 ISBN 978-3-032-08120-9 (eBook)
https://doi.org/10.1007/978-3-032-08120-9

This Springer imprint is published by the registered company Springer Nature Switzerland AG
The registered company address is: Gewerbestrasse 11, 6330 Cham, Switzerland

Preface

Digital Twin (DT) technology has rapidly emerged as a transformative paradigm, enabling the creation of virtual replicas of physical entities, processes, and systems that evolve in real time to mirror their real-world counterparts. By integrating advanced sensing, data analytics, and modeling techniques, DTs support predictive maintenance, system optimization, and enhanced decision-making across industries such as manufacturing, health care, and smart cities. The deployment of DTs over wireless networks, however, introduces unique challenges, including latency, reliability, limited communication resources, and the dynamic interplay between evolving systems and fluctuating network conditions. Addressing these challenges requires new frameworks that ensure high-quality DT construction, real-time service interaction, and adaptive evolution in dynamic wireless environments.

This book presents a comprehensive study of self-evolving digital twins over wireless networks, emphasizing the joint design of DT construction and service interaction under practical constraints. In particular, it explores key topics such as federated DT construction, quality-of-experience-driven service interaction under uncertain evolution, and adaptive management of network and computational resources. Innovative methodologies, including game-theoretic optimization, deep reinforcement learning, and generative AI-driven frameworks, are introduced to address real-world challenges in DT deployment and operation. Practical implementations, such as a 4D traffic scene generation framework and a video-assisted DT testbed for intelligent transportation systems, are provided to demonstrate the applicability of the proposed approaches. This book is organized as follows:

Chapter 1 introduces the concept, architecture, and features of DTs over wireless networks and discusses their benefits and representative application scenarios.

Chapter 2 identifies the key design requirements and challenges for DT modeling and wireless network integration and reviews enabling technologies that support high-fidelity DT construction and responsive service interaction.

Chapter 3 focuses on federated DT construction under dynamic evolution, formulating optimization problems and presenting a game-theoretic solution with two-sided matching and overlapping coalitions.

Chapter 4 addresses quality-of-experience-driven DT service interaction with uncertain evolution, introducing a prompt-guided decision transformer integrated with zero-forcing optimization to achieve real-time adaptive control.

Chapter 5 presents practical implementations of DTs in vehicular networks, including a surveillance video-assisted federated DT framework and an AI-driven real-time interactive 4D traffic scene generation system.

Chapter 6 concludes the book with a summary of key findings and outlines emerging research directions for self-evolving DTs over wireless networks.

We would like to thank all the SmilinNet group members for their valuable suggestions and comments. Special thanks go to the Series Editor, Prof. Xuemin (Sherman) Shen, and all the editorial staff at Springer Nature for their help throughout the publication process.

Nanjing, China Changyan Yi
Nanjing, China Ruoyang Chen
Nanjing, China Jiayuan Chen
Changsha, China Xiaolong Li
Montreal, QC, Canada Jun Cai

Competing Interests The authors have no competing interests to declare that are relevant to the content of this manuscript.

Contents

Acronyms

3DGS	3D Gaussian Splatting
ABAC	Attribute-Based Access Control
AES	Advanced Encryption Standard
AI	Artificial Intelligence
AIGC	AI-Generated Content
ANC	Active Noise Control
APIs	Application Programming Interfaces
AR	Augmented Reality
BIM	Building Information Modeling
CBS	Credit-Based Shaping
CEP	Complex Event Processing
CNNs	Convolutional Neural Networks
CRDT	Conflict-Free Replicated Data Type
CSI	Channel State Information
C-V2X	Cellular Vehicle-to-Everything
DDS	Data Distribution Service
DLT	Distributed Ledger Technology
DP	Differential Privacy
DRL	Deep Reinforcement Learning
DSRC	Dedicated Short-Range Communication
DVFS	Dynamic Voltage and Frequency Scaling
E2EE	End-to-End Encryption
EKF	Extended Kalman Filter
EPG	Exact Potential Game
ESs	Edge Servers
FDT	Federated Digital Twin
FPGAs	Field-Programmable Gate Arrays
GAN	Generative Adversarial Network
HFL	Hierarchical Federated Learning
IIoT	Industrial Internet of Things
IoT	Internet of Things

Chapter 1
Introduction to Digital Twin Over Wireless Networks

1.1 Overview of DT Over Wireless Networks

1.1.1 Concepts of DT Over Wireless Networks

Digital Twin (DT) has emerged as a transformative technology that creates high-fidelity, dynamically synchronized virtual counterparts of physical entities, fundamentally reshaping the interaction between the physical and digital worlds. First proposed in 2003 as a simulation-based concept aimed at replicating real-world physical entities [53], DT has since evolved into a multidisciplinary framework enabled by a convergence of advanced technologies, including big data, artificial intelligence (AI), 5G/6G, the Internet of Things (IoT), data visualization, extended reality (XR), and blockchain. As a cornerstone of Industry 4.0, DT has been widely applied in manufacturing [41], urban transportation [28], smart construction [13], and increasingly, in wireless communication systems [23]. A DT model serves as a continuously updated digital representation of a physical asset, reflecting both its current status and historical behavior to support predictive analysis, adaptive control, and optimized decision-making. While early DT applications primarily targeted static, non-living entities, modern DTs are capable of modeling dynamic, complex systems such as power plants, intelligent vehicles, or entire networks. In wireless networks, DTs enable real-time monitoring, fault prediction, and resource optimization by analyzing multi-source data from network infrastructure, user devices, and the surrounding environment. They also facilitate the simulation and testing of emerging communication technologies, ultimately enhancing the performance, scalability, and reliability of wireless systems.

Figure 1.1 illustrates the developmental timeline of DT technology. The term "Digital Twin" was first introduced in 2002 by Michael Grieves during a presentation on product lifecycle management. Shortly thereafter, Främling et al. [17] proposed an agent-based architecture to maintain virtual counterparts of physical products for real-time status monitoring [44], marking a shift from conceptual formulation

C. Yi et al., *Self-Evolving Digital Twin Over Wireless Networks*, Wireless Networks,
https://doi.org/10.1007/978-3-032-08120-9_1

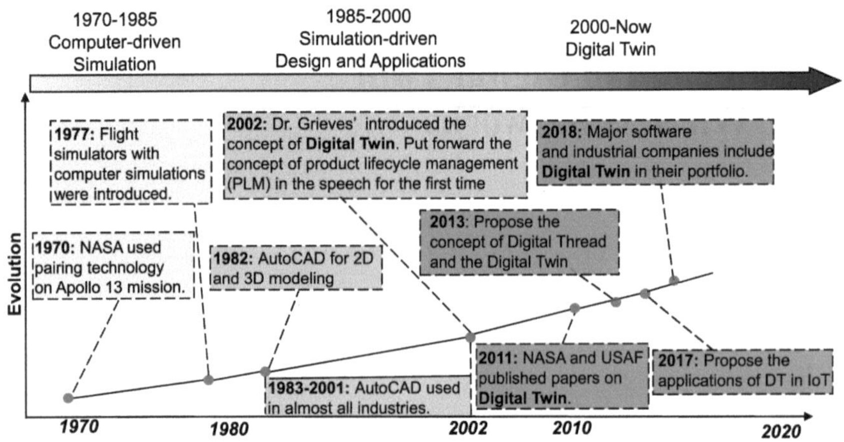

Fig. 1.1 The development timeline of general DT technology

to practical exploration. Prior to 2010, NASA operationalized the DT concept by deploying identical space vehicle models during the Apollo program to simulate flight conditions, thereby enabling effective training and maintenance, demonstrating DT's potential in aerospace and defense applications [42]. In 2017, Grieves formally defined the DT framework in a white paper [20], identifying its core components: physical entities, their virtual representations, and bidirectional data communication channels. Since then, DT technology has gained significant traction in industrial sectors, particularly in the Industrial Internet of Things (IIoT) and manufacturing, where it is widely used for predictive diagnostics, process optimization, and performance management [50]. Reflecting growing industry recognition, Gartner identified DT as one of the top 10 strategic technologies in 2017 and forecasted the deployment of millions of twin-enabled devices within five years [18, 40].

In recent years, DT has been extensively explored beyond manufacturing, with growing applications in aviation, healthcare, smart cities, intelligent transportation, urban planning, and next-generation wireless communications. At its core, DT enables the collection of both real-time and historical operational data from physical systems, facilitating predictive analysis, performance optimization, and intelligent decision-making. In aviation, DTs are utilized for aircraft maintenance, structural monitoring, and risk prediction. For instance, the authors in [5] proposed a DT model to detect fatigue cracks in aircraft wings. In healthcare, the integration of DT with wearable sensors and IoT devices allows for continuous monitoring of patients' physiological and treatment-related data, thus supporting more personalized and proactive medical care. In intelligent transportation systems, DTs enable dynamic traffic management, route planning, and infrastructure maintenance by mirroring real-time road conditions and system states. While DT applications in wireless communication are still in their early stages, several pioneering efforts have demonstrated its potential to enhance network performance. For example, Lu et al. [31] proposed a DT edge network architecture that integrates DT with blockchain to ensure secure and accurate edge computing in multi-access edge computing (MAEC) environments.

Recently, the integration of DT technology into wireless communication systems, i.e., DT over wireless network, has attracted growing attention [64]. Unlike conventional DT applications that focus solely on replicating physical entities, DT over wireless network not only involves the deployment, migration, and orchestration of DT services within wireless infrastructures, but also leverages the distributed nature of the wireleess networks to enable collaborative DT model construction and update across end devices, edge servers (ESs), and cloud platforms. This paradigm supports the joint utilization of heterogeneous wireless resources, reduces both the latency and energy consumption through localized processing, and facilitates adaptive, real-time DT model evolutions in dynamic communication environments.

Early research on DT over wireless network primarily addressed network modeling and state monitoring [55, 60, 69]. Wu et al. [58] provided a comprehensive overview of this field, outlining definitions, enabling technologies, challenges, and application scenarios. More recent studies have demonstrated DT's potential to optimize wireless system performance [22]. Dong et al. [15] pioneered its integration into MEC by proposing a deep learning-based framework for joint optimization of user association, resource allocation, and task offloading. Lu et al. [31] designed a DT model to mitigate unreliable and long-distance communications between end users and edge servers. van Huynh et al. [52] formulated a joint optimization problem to minimize end-to-end latency in DT-enabled ultra-reliable low-latency communication (URLLC) edge networks. Shi et al. [46] developed a DT-based wireless active noise control (ANC) system that offloads computationally intensive filtering to the cloud while maintaining real-time local noise cancellation. Khan et al. [22] presented a taxonomy of DT-enabled wireless systems, identifying design considerations and potential solutions, and later outlined architectures for DT over wireless network in 6G systems with edge-cloud integration and deployment trends [23]. Lu et al. [32] further proposed a deep reinforcement learning (DRL)-based optimization framework for adaptive edge association in 6G scenarios. Yaqoob et al. [63] introduced the concept as a key enabler for 6G network slicing, integrating AI for end-to-end Quality of Service (QoS) provisioning, while Tao et al. [51] proposed a hierarchical generative AI-enabled DT over wireless network framework for 6G. Li [25] analyzed DT technologies for 6G communications, proposing an end-to-end non-terrestrial network (NTN) architecture and discussing open research challenges. The overall development timeline of DT over wireless network is shown in Fig. 1.2.

1.1.2 Architecture of DT Over Wireless Networks

The architecture of DT over wireless networks consists of two major components: (i) the DT components to enable real-time monitoring, optimization, and autonomous operation; and (ii) the network components to ensure seamless communication, ubiquitous connectivity, and stable data exchange necessary for real-time synchroniza-

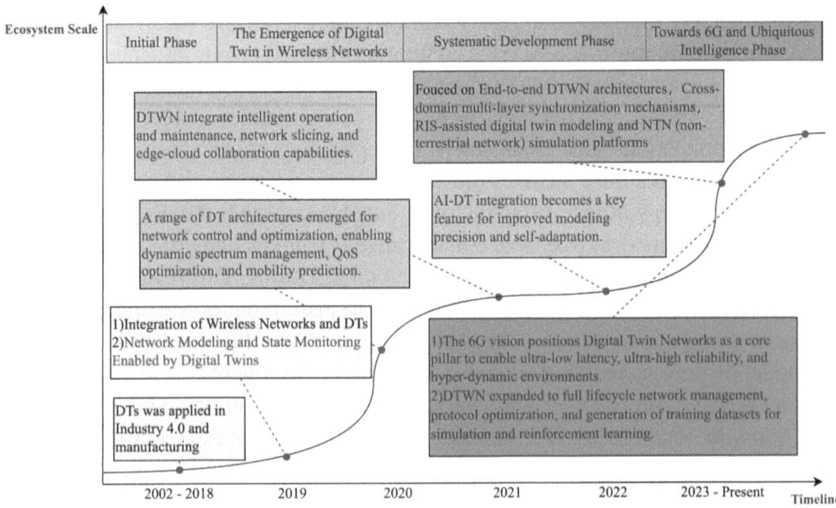

Fig. 1.2 The development timeline of DT over wireless networks

tion between the DT and the physical world [1]. These two components need to continuously interact to enable the ever-evolving DT construction, deployment, and application. Particularly, the DT components should be composed of multi-modal feature data, multi-functional model and highly-salable software that work together to replicate, monitor, and simulate the behavior of physical entities in a digital space. In the following, we will discuss the fundamental DT components in terms of the data layer, model layer, and software layer.

Data Layer: Multi-modal Data Construction and Management. The data layer serves as the foundation for the entire DT model. This layer is responsible for collecting and processing multi-modal data from various sources, such as sensors, cameras, radar systems, and network devices. In the context of wireless networks, the data layer handles a diverse range of data types that provide comprehensive insights into the system's state and performance.

- Multi-modal data collection: The data layer integrates multi-modal feature data collected from various data sources to create and update the DT models. For example, in a Vehicular Network (V2X), data collection encompasses vehicles, traffic cameras, and traffic signal systems, capturing real-time vehicle positions (trajectories), speeds, traffic flow information, and environmental factors like weather conditions and road status [21]. The integration of such diverse data sources yields a rich, dynamic dataset essential for constructing accurate DT representations.
- Data fusion and preprocessing: After data collection, the raw data should undergo preprocessing (including cleaning, noise reduction, and normalization) and advanced fusion techniques to ensure accuracy and consistency. These processes integrate information from heterogeneous sources into a unified, coherent view of the phys-

ical world. Preprocessing and fusion tasks could be performed at edge computing nodes or cloud platforms, depending on application requirements and underlying network capabilities.

Model Layer: The model layer is the core of realizing the accurate and predictive functions of DT. It processes the data collected from the data layer to construct a dynamic digital representation of the physical entity. This layer includes a variety of models designed to capture different aspects of the physical system's behavior. DT models in wireless networks are often multi-dimensional, encompassing aspects such as visual, geometric, rule-based, behavioral, and analytical characteristics.

- Visual models: Visual models provide realistic 3D virtual depictions of the physical system and its environment. These models are particularly valuable in domains like smart cities, traffic management, and industrial automation, enabling real-time visualization of system status. Technologies like Neural Radiance Fields (NeRFs) [54] and 3D Gaussian Splatting (3DGS) [24] are often used to generate high-fidelity 3D representations.
- Geometric models: Geometric models defines the physical structure and spatial configuration of objects within the system. In the context of vehicular networks, geometric models could include road layouts, intersection configurations, and the physical attributes of vehicles and infrastructure [14]. These models are essential for simulating traffic flow, vehicle navigation, and infrastructure design.
- Rule models: Rule-based models govern the DT's behavior based on predefined rules or professional knowledge. For example, in smart traffic management, rule models can control the operation of traffic lights, vehicle routing, and speed limits based on traffic density, road conditions, and time of day [27]. These models provide the necessary logic to govern decision-making in dynamic environments.
- Behavior models: Behavior models dynamically simulate the dynamic behavior of the physical entities. These models focus on how entities in the system interact with one another and evolve, capturing the changes based on the system's inputs and environmental factors. For example, in manufacturing systems, behavior models can simulate the operation of either the robots or human worker in a production line [59]. They would take into account factors such as both the robot's and human worker's task prioritization, response to faulty items on the line, or adjustments to varying production rates. These models are essential for understanding the real-time functioning and performance of automated systems.
- Data analytic models: Utilizing statistical analysis and machine learning techniques, these models are able to analyze the collected feature data and generate insights such as predictive analytics, fault detection, and performance optimization. They can help in identifying patterns and trends in the system's operation and provide valuable inputs for system improvements.

Software Layer: The software layer serves as the operational carrier of the DT, providing the interfaces necessary for deploying, managing, and utilizing DT models, thereby bridging the physical and digital environments. This layer facilitates interaction between system components and supports system optimization and decision-making.

- DT platforms: These platforms provide the infrastructure to deploy, manage, and orchestrate DT models across the system. They support services such as model deployment, updates, and coordination of data flows. These platforms also offer the ability to scale across different devices and network environments.
- Visualization and user interface: A key aspect of the software layer is the ability to visualize the system's state in real-time. This includes graphical representations of the DT, dashboards for monitoring system performance, and interactive interfaces for users to engage with the model. These tools are vital for decision-makers to understand and act on the insights provided by DT.
- APIs and interfaces: The software layer exposes standardized APIs (e.g., RESTful APIs) that allow other applications and systems to interact with DT. These interfaces enable seamless integration with network controllers, maintenance systems, and decision-making platforms. APIs facilitate the exchange of control messages between the physical and twin layers, ensuring synchronized operation.
- Security and privacy management: Ensuring data privacy and security is critical in any DT system. The software layer incorporates various security mechanisms, including authentication, authorization, and encryption, to protect sensitive data and ensure secure communication between the physical and digital layers.

The seamless interaction between these layers ensures that the virtual representation of the physical system stays synchronized and updated, reflecting changes in both the physical world and the underlying network infrastructure.

1.1.3 Features of DT Over Wireless Networks

The DT provides precise mapping from the physical world to digital space, with real-time interactions. DT mitigates the huge gap between physical space and digital space through continuous synchronization and updates. The features of DT over wireless networks can be summarized as follows (Fig. 1.3).

Precise Mapping: DT technology establishes an accurate and dynamic mapping between physical objects and their digital counterparts, serving as a bridge between the tangible and virtual worlds. By synchronizing historical data (such as design specifications, past performance metrics, and maintenance records) and real-time operational status (including sensor readings, environmental conditions, and system alerts) from the physical space to the digital space, DT models enable comprehensive data processing and analysis. This mapping must precisely reflect the full state of physical objects (e.g., geometric, functional, and behavioral attributes) with minimal

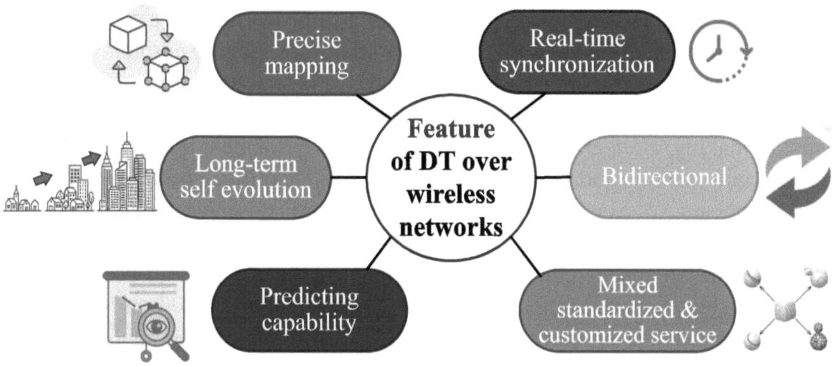

Fig. 1.3 Feature of DT over wireless networks

error, ensuring a high-fidelity virtual representation that supports advanced simulations, scenario testing, and optimizations.

The precision of this mapping is critical, as even minor discrepancies can lead to flawed insights or suboptimal decisions. For example, in industrial settings, a DT model of a manufacturing assembly line must account for mechanical wear, thermal expansion, and vibrational dynamics to accurately predict failures or bottlenecks [16]. Advanced techniques like 3D modeling, computational physics, and probabilistic simulations are employed to enhance the twin's accuracy. Furthermore, the integration of wireless networks introduces challenges such as data packet loss or latency, necessitating robust error-correction mechanisms and adaptive sampling rates to preserve mapping fidelity.

Real-Time Synchronization: Unlike traditional simulation methods, which often rely on static or periodic updates, DT models naturally maintain continuous, real-time synchronization with physical objects. Data collected from distributed sensors, IoT devices, and control systems are instantly processed on the digital side to update the twin's model, ensuring it mirrors the current state of its physical counterpart. To achieve this, edge computing plays a pivotal role by decentralizing computational tasks, thereby minimizing latency and ensuring timely feedback.

For instance, in a smart city deployment, traffic flow sensors transmit data to edge servers, which preprocess and forward only the most relevant information to the cloud-based DT [7]. This hierarchical data handling reduces bandwidth consumption and delays, enabling the twin to dynamically adapt to changes, such as rerouting vehicles in response to an accident. Real-time synchronization also demands stringent QoS guarantees from wireless networks, including high reliability, low jitter, and prioritized data streams for critical updates. Technologies like 5G URLLC [56] and Time-Sensitive Networking (TSN) [33] are instrumental in meeting these requirements.

Long-Term Self-evolution: DT models are designed to evolve over time by continuously learning from accumulated data, adapting to new operational patterns, and refining their predictive models. Through machine learning (e.g., reinforcement learning, neural networks) and AI-driven analytics, they identify correlations, detect anomalies, and improve their accuracy in forecasting system behaviors. This self-optimization ensures that the DT models remain relevant and effective in supporting decision-making, even as the physical system or environment undergoes changes, such as equipment degradation, seasonal variations, or shifts in user demand.

In healthcare, for example, a DT of a patient's cardiovascular system might integrate data from wearable devices, electronic health records, and genetic profiles to personalize treatment plans [8]. Over time, the twin learns to predict health risks with greater precision, adjusting its recommendations based on longitudinal data. However, self-evolution also raises challenges in model governance, requiring mechanisms to prevent bias drift, ensure explainability, and maintain alignment with physical-world constraints.

Bidirectional Interaction: A key feature of DT is its bidirectional interaction with physical objects, forming a closed-loop cyber-physical system. While physical entities constantly feed real-time data into their digital counterparts, the twins, in turn, provide actionable insights (e.g., predictive maintenance alerts), optimizations (e.g., energy-efficient control strategies), and even direct control commands (e.g., adjusting robotic arm trajectories) back to the physical world. This feedback mechanism enhances system efficiency, enabling adaptive and intelligent operations that are responsive to dynamic conditions.

In agriculture, soil sensors relay moisture and nutrient levels to a crop's DT, which analyzes the data alongside weather forecasts to optimize irrigation schedules [4]. Commands are then sent to automated sprinklers, reducing water waste while maximizing yield. Bidirectional interaction relies heavily on secure, low-latency wireless communication, with protocols like MQTT and CoAP ensuring efficient data exchange. Additionally, safety-critical applications (e.g., autonomous vehicles) demand fail-safe mechanisms to prevent erroneous commands during network disruptions.

Predictive Capability: Beyond real-time monitoring, DT leverages AI and historical data to forecast future states of physical systems, transforming reactive operations into proactive strategies. In smart transportation, for instance, DT analyzes traffic patterns, weather conditions, and event schedules to predict congestion hotspots and recommend optimal routes, reducing commute times [65]. Similarly, in energy grids, twins simulate demand surges and renewable generation fluctuations to preemptively balance loads and prevent blackouts [3].

Predictive accuracy hinges on the twin's ability to fuse multi-source data (e.g., sensor feeds, maintenance logs, external APIs) and employ advanced algorithms like digital thread analysis or federated learning. Wireless networks must support these compute-intensive tasks by enabling seamless data aggregation from heterogeneous devices, often across wide areas. Edge AI can further decentralize prediction workloads, allowing localized decision-making without overwhelming central servers.

Mixed Standardized and Customized Service: DT platforms offer a hybrid approach, combining standardized frameworks with customizable modules to meet diverse industry needs. Core functionalities, such as data ingestion (via protocols like OPC UA), real-time processing (using stream analytics engines), and visualization (through unified dashboards), follow universal standards to ensure interoperability and scalability. Meanwhile, domain-specific applications (e.g., predictive maintenance in manufacturing, smart grid optimization, or patient-specific healthcare modeling) allow tailored solutions that address unique challenges.

For example, a standardized DT platform for industrial equipment might include generic APIs for condition monitoring [6], while a customized extension could integrate proprietary algorithms for specific applications. Wireless networks facilitate this flexibility by supporting modular architectures, such as microservices or containerized deployments [47–49], which can be dynamically reconfigured via software-defined networking (SDN). This adaptability ensures DT models can scale across use cases from single machines to city-wide infrastructures without sacrificing performance or compatibility.

1.1.4 Benefits of DT Over Wireless Networks

DTs integrated with wireless networks are revolutionizing how we interact with and optimize physical systems through three fundamental processes: optimization, prediction, and automation. This powerful combination creates a dynamic, real-time bridge between the physical and digital worlds, enabling unprecedented levels of system intelligence and responsiveness. The wireless connectivity layer, including 5G/6G, Wi-Fi 6, Low-Power Wide-Area Network (LPWAN), and emerging technologies, serves as the critical nervous system that breathes life into DT implementations across industries. Below we explore six key benefits that this integration delivers, each representing a quantum leap in system management and operational efficiency (Fig. 1.4).

High-Fidelity Virtual Representation: The foundation of any effective DT lies in its ability to create an exceptionally accurate virtual counterpart that precisely mirrors the physical entity's structural, functional, and behavioral characteristics. Modern high-fidelity DTs achieve this through multi-physics modeling techniques that simulate mechanical, thermal, electrical, and fluid dynamics interactions simultaneously. Wireless networks enable this fidelity by supporting massive IoT deployments where hundreds of sensors, e.g., high-resolution Light Detection And Ranging (LiDAR) units capturing spatial details to Micro-Electro-Mechanical System (MEMS) accelerometers tracking microscopic vibrations, continuously feed data to the twin. For instance, in precision manufacturing, wireless-enabled force-torque sensors on robotic arms provide real-time telemetry at sub-millimeter accuracy, allowing the DT to detect tool wear patterns invisible to human eyes [34, 36]. The emergence of millimeter-wave wireless technologies (like 5G NR) with their multi-gigabit

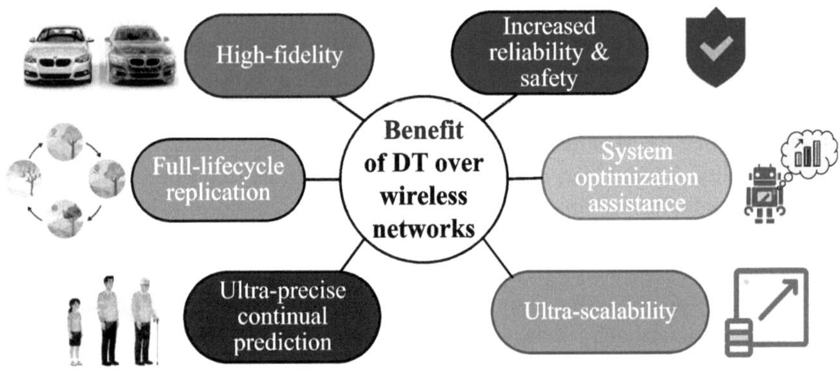

Fig. 1.4 Benefit of DT over wireless networks

throughput capabilities makes possible the transmission of ultra-high-definition 3D point clouds that form the basis of photorealistic DTs. This level of detail enables engineers to perform virtual stress tests, thermal analyses, and aerodynamic simulations with confidence that the results will perfectly correlate with physical world behavior.

Full-Lifecycle Replication: DTs truly distinguish themselves through their ability to accompany a physical asset throughout its entire existence from conceptual design through to decommissioning. Wireless networks facilitate this continuous lifecycle tracking by providing ubiquitous connectivity regardless of the asset's location or mobility status. During the design phase, wireless-connected augmented reality systems allow engineers to overlay DT simulations onto physical prototypes for instant validation. In operational phases, wireless sensor networks monitor equipment health indicators like vibration spectra, thermal profiles, and lubrication quality, feeding this data to predictive maintenance algorithms. For mobile assets like commercial aircraft, satellite and air-to-ground wireless links ensure the DT receives updates even during transoceanic flights. As the asset ages, the DT accumulates a comprehensive "health journal" that informs life-extension decisions. A compelling example is modern wind farms, where each turbine's DT analyzes its load history (transmitted via hybrid cellular-satellite links) to recommend component replacements before failures occur, effectively doubling the operational lifespan of these multimillion-dollar assets while maintaining safety margins.

Ultra-Precise Continual Prediction: The predictive power of wireless-enabled DTs represents a paradigm shift from traditional maintenance approaches. By combining historical performance data with real-time sensor feeds transmitted over low-latency wireless connections, these systems can forecast future system states with remarkable accuracy. Advanced implementations employ federated learning architectures where edge devices preprocess sensor data locally (reducing wireless bandwidth needs) while contributing to continuously improving prediction models [10, 62]. In smart city applications, DTs analyze traffic patterns from thousands of wireless-connected cameras and induction loops, predicting congestion in advance. Industrial implementations go further, for example, semiconductor fabrication plants use wireless vibration monitoring paired with DTs to predict equipment failures in advance, scheduling maintenance during natural production breaks. The predictive capability extends to financial optimization as well; logistics companies use DTs to simulate fleet operations under various weather and demand scenarios transmitted via 5G networks, enabling dynamic pricing and efficient resource allocation.

Increased Reliability and Safety: In safety-critical domains like transportation and energy infrastructure, wireless-enabled DTs create an always-on protective shield around physical systems. Vehicle DTs in intelligent transportation systems process data from hundreds of onboard sensors transmitted via Vehicle-to-Everything (V2X) networks, maintaining a 360° awareness of the vehicle's condition and surroundings. When the twin detects abnormal brake temperature patterns or tire wear characteristics, it can automatically schedule maintenance before safety margins are compromised. More dramatically, in chemical plants, wireless gas sensors and thermal cameras feed data to process DTs that run real-time explosion risk simulations; if dangerous conditions are predicted, the system can automatically initiate shutdown procedures within milliseconds via wireless control networks. The aviation industry provides perhaps the most rigorous implementation, for example, modern aircraft maintain DTs that cross-validate data from thousands of sensors via redundant wireless data buses, enabling the detection of potential system failures long before they become critical. This capability has contributed to the remarkable safety record of modern aviation, with DTs now considered essential for next-generation autonomous urban air mobility systems.

System Optimization Assistance: DTs connected via high-performance wireless networks serve as tireless optimization engines for complex systems. In manufacturing facilities, wireless-enabled DTs continuously analyze the interplay between machine performance, environmental conditions, and production demands to recommend optimal operating parameters. For example, in automotive assembly plants, DTs can use real-time data from wireless torque wrenches and vision systems to dynamically adjust assembly line speeds, reducing energy consumption while maintaining throughput [66]. Building management systems take this further, for example, wireless sensors monitoring occupancy, weather, and equipment performance feed data to building DTs that optimize air conditioner operation on a per-zone basis, achieving energy savings in smart campuses. The most advanced implementations employ closed-loop optimization where the DT's recommendations are automatically implemented via wireless control systems, creating self-optimizing facilities that continuously tune their own performance. Retail chains are now applying this concept, using DTs to optimize store layouts and inventory placement based on wireless tracking of customer movement patterns.

Ultra-Scalability: The true power of wireless-enabled DTs emerges when scaling from individual assets to entire ecosystems. Modern implementations leverage hierarchical architectures where device-level twins communicate with system-level twins via 5G or Wi-Fi 6, which in turn integrate into enterprise-wide DT networks. Cloud-edge computing paradigms enabled by high-bandwidth wireless backhauls allow this scaling without overwhelming central systems. Singapore's Virtual Singapore project exemplifies this, combining DTs of buildings (connected via LPWAN), transportation networks (5G-enabled), and utility systems into a comprehensive urban simulation platform. Wireless network slicing ensures critical twin data (like emergency service coordination) receives priority during network congestion. The scalability extends vertically as well, for instance, a single factory machine's DT might process local sensor data via Bluetooth mesh networks, while contributing to a global twin analyzing production patterns across dozens of factories worldwide via satellite links. This scalability is driving adoption in global supply chains, where wireless-connected DTs monitor container conditions, predict shipping delays, and optimize routes across multimodal transportation networks, reducing logistics costs for early adopters.

1.2 Application Scenarios

As a central catalyst in the ongoing digital transformation, DT technology is achieving revolutionary breakthroughs through its deep integration with wireless communication systems. The continued maturity of 5G deployment and the intensification of 6G research are laying the groundwork for DTs with communication capabilities characterized by ultra-low latency, exceptional reliability, and ubiquitous coverage across diverse environments. This synergy is enabling DTs to move beyond static equipment models and toward dynamic simulations of complex systems. Such evolution has substantial applications in critical domains, including smart manufacturing, intelligent transportation systems, and remote healthcare. This book provides a systematic analysis of how wireless communication technologies underpin DT frameworks and explores their innovative applications in four key areas: 6G-based network environments, manufacturing systems enhanced by wireless DTs, transportation networks empowered by digital twin representations, and remote healthcare delivered via wireless DT platforms (Fig. 1.5).

6G Network Access and Control: As the next-generation communication infrastructure expected for 2030, the technological innovation of the 6G network provides unprecedented technical support for DTs [2]. As the likely successor to 5G anticipated around 2030, 6G is expected to underpin DT applications through breakthroughs in semantic communication, edge-based deterministic control, dynamic network slicing, and energy-efficient operation. In wireless access, semantic communication techniques focus on transmitting compact, task-relevant representations, such as pose landmarks, semantic segmentation maps, or abstract embeddings, instead of full raw media, with generative AI on the receiver side used to reconstruct or interpret the needed content. This approach offers substantial efficiency gains over conventional transmission. On the edge side, integrating wireless systems with time-sensitive net-

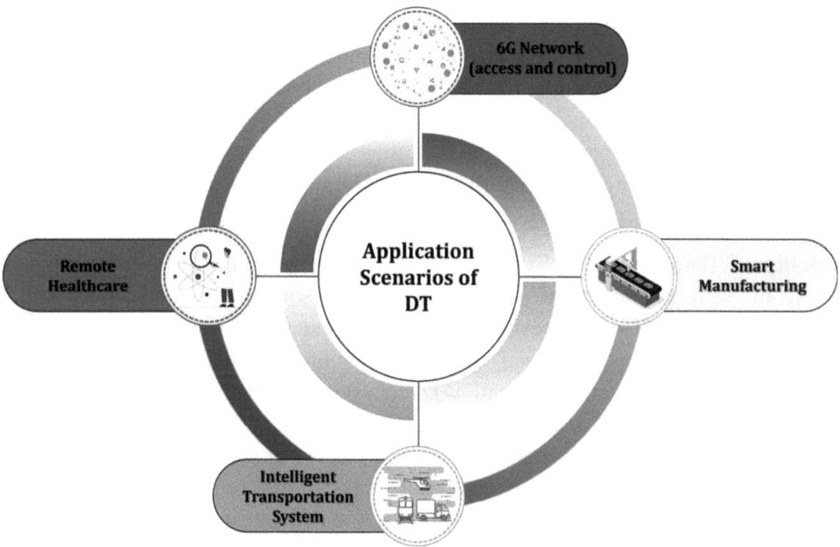

Fig. 1.5 Application scenarios of DT over wireless networks

working (TSN) supports prioritized, low-latency traffic handling–particularly for haptic or control signals–and can enable single-digit millisecond end-to-end delays in controlled or private deployments. For multimodal DT scenarios combining high-resolution video and haptic feedback, event-driven or deadband sampling strategies help reduce unnecessary data transmissions, athough precise synchronization remains a technical challenge. Meanwhile, AI-driven network slicing and dynamic quality-of-service orchestration can adjust latency, reliability, and bandwidth allocation in real time to meet diverse DT service requirements, subject to infrastructure maturity. Finally, energy efficiency, through adaptive carrier management, sleep scheduling, and slicing-aware optimization, is a main objective, with typical studies showing double-digit percentage savings in radio access network energy use.

Smart Manufacturing: In the field of smart manufacturing, wireless DT technology is driving the transformation from Industry 4.0 to human-centered Industry 5.0 [37–39]. In the realm of smart manufacturing, wireless-enabled DT technologies are poised to accelerate the shift from Industry 4.0 toward human-centric Industry 5.0, by enabling real-time virtual replicas and process optimization. In manufacturing settings, DT systems can ingest data from distributed wireless sensors, such as EMG bands, RFID tags, torque or motion sensors, or even wearable visual devices, to model operator ergonomics, tool usage patterns, or assembly dynamics. These models enable adaptive control feedback, ergonomic guidance, and predictive maintenance without physical disruptions. Edge-deployed inference engines process data with millisecond-level latency, supporting responsive anomaly detection and intervention. For instance, DT-assisted anomaly detection using federated learning methods can improve fault response while preserving data privacy across distributed units.

Wireless networks, especially private 5G setups that combine uRLLC and mMTC features, offer low-latency, high-reliability channels to synchronize DT inputs and real-world actions. Meanwhile, intelligent slicing and AI-driven orchestration ensure that critical control streams remain prioritized, and energy-aware mechanisms help maintain sustainable operations. Together, these capabilities make wireless DTs an enabling backbone for responsive, efficient, and human-aligned smart manufacturing systems.

Intelligent Transportation System: Intelligent transportation systems are upgrading from "vehicle-road-cloud" collaboration to "person-vehicle-road-environment" full-element DTs [19]. The evolution of intelligent transportation systems (ITS) is moving beyond the traditional 'vehicle–road–cloud' model toward an inclusive 'person–vehicle–road–environment' DT ecosystem that captures human behavior, vehicular dynamics, infrastructure, and environmental context. In this vision, DT frameworks ingest real-time data streams, such as driver state monitoring, environmental perception from cameras and radar, traffic signal timing, and road surface conditions through wireless V2X communications. Edge-deployed DT processors synthesize this multimodal information to anticipate safety-critical events, support adaptive signal control, and enable proactive warnings. Cellular V2X (C-V2X) and related private 5G technologies provide the low-latency and high-reliability communication channels essential for synchronizing virtual and physical twins in mobility systems. Meanwhile, AI-powered DT engines enhance scene generation and simulation for scenario-based validation of autonomous driving strategies. Smart city deployments leverage DT-informed traffic optimization and emission-aware path planning to improve flow efficiency and environmental performance. Together, these wireless DT-enabled capabilities promise a more proactive, context-aware, and resilient transportation ecosystem.

Remote Healthcare: In the field of telemedicine, wireless DT technology is reshaping the diagnosis and treatment model [35]. Wireless-enabled DT systems are redefining the landscape of remote healthcare by enabling continuously updated, individualized virtual representations of patients. Such systems integrate data from diverse sources–wearable sensors capturing vital signs, imaging devices delivering anatomical context, environmental monitors reporting ambient variables, and mobile apps gathering behavioral inputs, to maintain a real-time digital reflection of a patient's health profile. These DTs operate at the edge of the network, allowing rapid detection of physiological anomalies, real-time adaptation of treatment recommendations, and preemptive simulation of clinical scenarios without patient mobility constraints. Key enabling technologies include ultra-reliable, low-latency wireless communication, such as private 5G (URLLC) networks, that synchronize data streams and trigger timely alerts and interventions. By employing federated learning, DT models can refine their predictive accuracy across distributed healthcare sites while preserving privacy and adhering to data governance frameworks. Edge and fog computing architectures complement this by reducing dependence on centralized cloud infrastructure, thus mitigating latency and enhancing data security. Moreover, DT-based simulations support virtual rehearsal of complex procedures, risk stratification for personalized therapy planning, and scalable telemedicine workflows that reduce inpatient load and extend specialist access to remote or underserved communities. In combination, these technologies cultivate a responsive, efficient, and patient-centered paradigm for delivering resilient and equitable remote healthcare.

1.3 DT Construction and Service Interaction Over Wireless Networks

In DT over wireless networks, two fundamental processes must be jointly addressed to fully realize its potential: (i) constructing and update high-fidelity DT models via distributed sensing and edge intelligence; and (ii) enabling efficient service interaction between the physical entities and their virtual counterparts. The former ensures that DTs faithfully reflect the evolving state of physical entities through continuous data acquisition, processing, state synchronization and model training, while the latter allows users and applications to query, control, and exploit DT capabilities in real time. These processes are inherently intertwined and subject to the limitations of wireless communications, such as constrained bandwidth, latency, and reliability, which necessitate a careful co-design. The following subsections elaborate on these two aspects in detail.

1.3.1 DT Construction Over Wireless Networks

DT over wireless networks requires an efficient and adaptive model construction process to accurately mirror the ever-changing state of physical entities [9, 10]. Traditional DT construction often relies on centralized processing, where sensor data is transmitted to a central cloud for model building and rendering. However, this approach faces significant challenges in wireless environments, including high communication overhead, processing delays, and the presence of data silos that hinder timely and comprehensive data collection. To address these issues, a distributed edge-cloud collaborative framework should be explored. As shown in Fig. 1.6, in this framework, the DT is constructed as a global model consisting of multiple functional components, or partial-DTs, each created at geographically distributed ESs using locally collected feature data. The central cloud coordinates this process by assigning partial-DTs to specific ESs, initializing core parameters, and later integrating the completed partial-DTs into a coherent global model. This design allows parallel and heterogeneous DT construction, where different partial-DTs may represent distinct aspects of the physical environment, such as vehicles, pedestrians, and infrastructure in a transportation network [26, 43], while still being synchronized as a whole. Furthermore, since the wireless network conditions and physical environments evolve over time [12, 29, 68], DT construction must also be dynamic, with partial-DTs

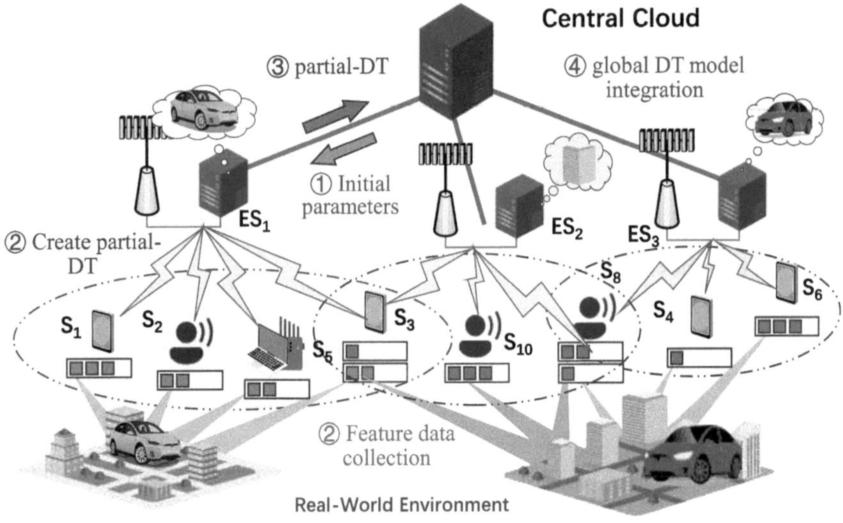

Fig. 1.6 System architecture of DT construction over wireless networks

continuously updated to maintain fidelity. Using the distributed nature of wireless networks, this approach not only reduces latency and bandwidth consumption, but also enhances scalability and adaptability, enabling high-quality DT services for applications such as intelligent transportation, smart manufacturing, and large-scale IoT systems [45, 57, 61].

1.3.2 DT Service Interaction Over Wireless Networks

DT service interaction over wireless networks refers to the bidirectional engagement between users and DT models through wireless links, allowing real-time monitoring, predictive analytics, and interactive control of physical entities. Unlike traditional wireless services, DT interactions require both uplink and downlink communications: users transmit high-dimensional interaction signals such as motions, voice, and facial expressions, while DT models provide immediate, customized feedback in forms such as virtual 3D simulations, multimedia responses, or analytical insights. The system architecture typically comprises a DT server hosting high-fidelity interactive models, wireless users with mobile devices, and enabling infrastructure such as reconfigurable intelligent surfaces (RIS) and edge servers to assist signal transmission, reduce latency, and enhance connectivity under dynamic wireless conditions. This design allows DT services to be accessible ubiquitously and responsively, while preserving the authenticity and accuracy of interactions. The key objectives of the interaction of the DT service include maximizing the quality of experience of users, which includes objective metrics such as latency and throughput, and subjective fac-

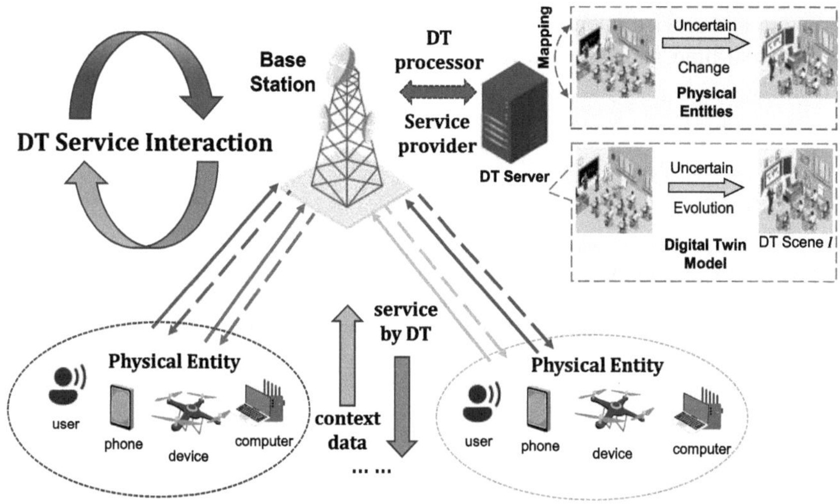

Fig. 1.7 System architecture of DT service interaction over wireless networks

tors such as perceived realism and responsiveness. Optimization variables involve resource allocation for uplink and downlink links [11], RIS phase adjustments [67], server beamforming [30], feedback signal resolution, and computation resource distribution, all coordinated to maintain seamless, high-quality, and adaptive interactions as DT scenes evolve over time (Fig. 1.7).

1.4 Summary of the Chapter

In this chapter, we have presented a comprehensive overview of DT over wireless networks, including its concept, architecture, key features, and benefits, while highlighting its distinctions from conventional simulation models and general DT services. We introduced the fundamental concept of DT over wireless networks and clarified its distinctions from conventional DT models. We summarized how DTs leverage wireless connectivity to achieve precise mapping, real-time synchronization, long-term self-evolution, predictive capabilities, and interactive feedback, thereby moving beyond static simulations to dynamic, adaptive, and intelligent representations of physical systems. We also outlined the architecture of DT over wireless networks, focusing on the data, model, and software layers, which jointly support scalable and secure DT construction and operation.

Furthermore, we highlighted the benefits and application scenarios of DT in smart manufacturing, intelligent transportation, and remote healthcare, demonstrating its potential to optimize performance, enhance safety, and enable automation across critical domains. Finally, we emphasized two essential enablers–distributed DT construction and service interaction–which provide the foundation for realizing high-fidelity, real-time, and user-centric DT services over next-generation wireless infrastructures.

References

1. Afzal, M.K., Naeem, M., Ejaz, W.: Digital Twins for Wireless Networks. Springer (2024)
2. Ahmadi, H., Nag, A., Khar, Z., Sayrafian, K., Rahardja, S.: Networked twins and twins of networks: an overview on the relationship between digital twins and 6G. IEEE Commun. Stand. Mag. **5**(4), 154–160 (2022)
3. Al-Shetwi, A.Q., Atawi, I.E., El-Hameed, M.A., Abuelrub, A.: Digital twin technology for renewable energy, smart grids, energy storage and vehicle-to-grid integration: advancements, applications, key players, challenges and future perspectives in modernising sustainable grids. IET Smart Grid **8**(1), e70026 (2025)
4. Alves, R.G., Maia, R.F., Lima, F.: Development of a digital twin for smart farming: Irrigation management system for water saving. J. Clean. Prod. **388**, 135920 (2023)
5. Bielefeldt, B., Hochhalter, J., Hartl, D.: Computationally efficient analysis of SMA sensory particles embedded in complex aerostructures using a substructure approach. In: Smart Materials, Adaptive Structures and Intelligent Systems, vol. 57298, p. V001T02A007. American Society of Mechanical Engineers (2015)
6. Broo, D.G., Bravo-Haro, M., Schooling, J.: Design and implementation of a smart infrastructure digital twin. Autom. Constr. **136**, 104171 (2022)
7. Campolo, C., Genovese, G., Molinaro, A., Pizzimenti, B., Ruggeri, G., Zappalà, D.M.: An edge-based digital twin framework for connected and autonomous vehicles: design and evaluation. IEEE Access **12**, 46290–46303 (2024)
8. Chen, J., Yi, C., Okegbile, S.D., Cai, J., Shen, X.S.: Networking architecture and key supporting technologies for human digital twin in personalized healthcare: a comprehensive survey. IEEE Commun. Surv. Tutor. **26**(1), 706–746 (2024)
9. Chen, R., Yi, C.: A game-theoretic online optimization for federated digital twin construction via wireless sensing. In: Proceedings of IEEE ICC, Montreal, Canada (2025)
10. Chen, R., Yi, C., Zhou, F., Kang, J., Wu, Y., Niyato, D.: Federated digital twin construction via distributed sensing: a game-theoretic online optimization with overlapping coalitions. IEEE Trans. Mob. Comput., 1–17 (2025)
11. Chen, R., Yi, C., Zhu, K., Cai, J., Chen, B.: A DRL-based hierarchical game for physical layer security with dynamic trilateral coalitions. In: Proceedings of IEEE ICC, pp. 4495–4500 (2023)
12. Chen, R., Yi, C., Zhu, K., Chen, B., Cai, J., Guizani, M.: A three-party hierarchical game for physical layer security aware wireless communications with dynamic trilateral coalitions. IEEE Trans. Wirel. Commun. **23**(5), 4815–4829 (2024)
13. Dang, H.V., Tatipamula, M., Nguyen, H.X.: Cloud-based digital twinning for structural health monitoring using deep learning. IEEE Trans Ind. Inform. **18**(6), 3820–3830 (2021)
14. Davletshina, D., Reja, V.K., Brilakis, I.: Automating construction of road digital twin geometry using context and location aware segmentation. Autom. Const. **168**, 105795 (2024)
15. Dong, R., She, C., Hardjawana, W., Li, Y., Vucetic, B.: Deep learning for hybrid 5G services in mobile edge computing systems: learn from a digital twin. IEEE Trans. Wirel. Commun. **18**(10), 4692–4707 (2019)
16. Feng, K., Xiao, H., Zhang, J., Ni, Q.: A digital twin methodology for vibration-based monitoring and prediction of gear wear. Wear, 205806 (2025)
17. Främling, K., Holmström, J., Ala-Risku, T., Kärkkäinen, M.: Product agents for handling information about physical objects. Report of Laboratory of Information Processing Science Series B, TKO-B, vol. 153, no. 03 (2003)

18. Gartner: Gartner's Top. 10 Strategic Technology Trends for 2018. Gartner, Stamford, CT, USA (2017)
19. Ge, C., Qin, S.: Digital twin intelligent transportation system (DT-ITS)—a systematic review. IET Intell. Transp. Syst. **18**(12), 2325–2358 (2024)
20. Grieves, M., Vickers, J.: Digital twin: mitigating unpredictable, undesirable emergent behavior in complex systems. In: Transdisciplinary Perspectives on Complex Systems: New Findings and Approaches, pp. 85–113. Springer (2016)
21. Irfan, M.S., Dasgupta, S., Rahman, M.: Toward transportation digital twin systems for traffic safety and mobility: a review. IEEE Internet Things J. **11**(14), 24581–24603 (2024)
22. Khan, L.U., Han, Z., Saad, W., Hossain, E., Guizani, M., Hong, C.S.: Digital twin of wireless systems: overview, taxonomy, challenges, and opportunities. IEEE Commun. Surv. Tutor. **24**(4), 2230–2254 (2022)
23. Khan, L.U., Saad, W., Niyato, D., Han, Z., Hong, C.S.: Digital-twin-enabled 6G: vision, architectural trends, and future directions. IEEE Commun. Mag. **60**(1), 74–80 (2022)
24. Khan, M.G., Guarese, R., Johnson, F., Wang, X.V., Bergman, A., Edvinsson, B., Romero, M., Vachier, J., Kronqvist, J.: PerfCam: digital twinning for production lines using 3D Gaussian splatting and vision models. IEEE Access (2025)
25. Li, L.: Research on digital twin technology for future 6G. Intell. Converg. Netw. **6**(1), 20–40 (2025)
26. Li, X., Wei, J., Wang, H., Dong, L., Chen, R., Yi, C., Cai, J., Niyato, D., Shen, X.: Towards intelligent transportation with pedestrians and vehicles in-the-loop: a surveillance video-assisted federated digital twin framework. IEEE Netw. pp. 1–1 (2025)
27. Liao, X., Leng, S., Sun, Y., Zhang, K., Imran, M.A.: A digital twin-based traffic guidance scheme for autonomous driving. IEEE Internet J. (2024)
28. Liao, X., Wang, Z., Zhao, X., Han, K., Tiwari, P., Barth, M.J., Wu, G.: Cooperative ramp merging design and field implementation: a digital twin approach based on vehicle-to-cloud communication. IEEE Trans. Intell. Transp. Syst. **23**(5), 4490–4500 (2021)
29. Liu, D., Chen, R., Yi, C., Zhang, T., Li, X., Li, J., Wang, R., Zhu, K.: A hierarchical game for physical layer security aware cooperative communications with dynamic interchangeable relaying and jamming. IEEE Trans. Veh. Technol. **74**(1), 968–983 (2025)
30. Liu, D., Chen, R., Zhang, T., Yi, C.: A DRL-based hierarchical game for physical layer security aware cooperative communications. In: Proceedings of IEEE ICCC, pp. 1769–1774 (2024)
31. Lu, Y., Huang, X., Zhang, K., Maharjan, S., Zhang, Y.: Low-latency federated learning and blockchain for edge association in digital twin empowered 6G networks. IEEE Trans Ind. Inform. **17**(7), 5098–5107 (2020)
32. Lu, Y., Maharjan, S., Zhang, Y.: Adaptive edge association for wireless digital twin networks in 6G. IEEE Internet Things J. **8**(22), 16219–16230 (2021)
33. Lu, Y., Zhao, G., Xu, C., Imran, M., Yu, K., Rodrigues, J.J.: A framework for digital twin-based deterministic communication in satellite time sensitive networks. In: Proceedings of IEEE ICC, pp. 6301–6306. IEEE (2023)
34. Man, T., Zhu, H., Chen, R., Yi, C.: Hierarchical DRL-based multi-motor control with torque synchronization in industrial IoT. In: International Conference on Wireless Artificial Intelligent Computing Systems and Applications (WASA), pp. 11–21. Springer (2025)
35. Mokhtari, M.: Human digital twins in personalized healthcare: an overview and future perspectives (2025). arXiv preprint arXiv:2503.11944
36. Muthuswamy, P., K, S.: Artificial intelligence based tool condition monitoring for digital twins and industry 4.0 applications. Int. J. Interact. Des. Manuf. (IJIDeM) **17**(3), 1067–1087 (2023)
37. Okegbile, S.D., Cai, J., Chen, J., Yi, C.: A reputation-enhanced shard-based byzantine fault-tolerant scheme for secure data sharing in zero trust human digital twin systems. IEEE Internet Things J. **11**(12), 22726–22741 (2024)
38. Okegbile, S.D., Cai, J., Yi, C., Niyato, D.: Human digital twin for personalized healthcare: vision, architecture and future directions. IEEE Netw. **37**(2), 262–269 (2023)
39. Okegbile, S.D., Cai, J., Zheng, H., Chen, J., Yi, C.: Differentially private federated multi-task learning framework for enhancing human-to-virtual connectivity in human digital twin. IEEE J. Sel. Areas Commun. **41**(11), 3533–3547 (2023)

40. Panetta, K.: Gartner Top 10. Strategic Technology Trends for 2019. Gartner, Stamford, CT, USA (2018)
41. Peng, K., Huang, H., Bilal, M., Xu, X.: Distributed incentives for intelligent offloading and resource allocation in digital twin driven smart industry. IEEE Trans Ind. Inform. **19**(3), 3133–3143 (2022)
42. Picazo-Vela, S., Hernández, L.R.: Technology, science and culture—a global vision. In: Technology, Science and Culture—A Global Vision, vol. II. IntechOpen (2020)
43. Qian, Z., Chen, R., Yi, C., Zhai, X., Chen, B.: Collision avoidance control for autonomous driving with multiple dynamic obstacles in IoV: a prediction-enhanced APF-based approach. IEEE Internet Things J. (2025)
44. Shafto, M., Conroy, M., Doyle, R., Glaessgen, E., Kemp, C., LeMoigne, J., Wang, L.: Draft modeling, simulation, information technology & processing roadmap. Technol. Area **11**, 1–32 (2010)
45. Shang, S., Yi, C., Zhang, T., Chen, R., Cai, J.: Latency-energy aware dynamic application placement for edge computing: a vacation queue based optimization approach. IEEE Trans. Netw. Sci. Eng. **11**(2), 2249–2263 (2024)
46. Shi, C., Du, F., Wu, Q.: A digital twin architecture for wireless networked adaptive active noise control. IEEE/ACM Trans. Audio Speech Lang. Process. **30**, 2768–2777 (2022)
47. Shi, Y., Wang, J., Yi, C., Wang, R., Chen, B.: Proactive application deployment for MEC: an adaptive optimization based on imperfect multi-dimensional prediction. IEEE Trans. Veh, Technol (2025)
48. Shi, Y., Yang, Y., Yi, C., Chen, B., Cai, J.: Towards online reliability-enhanced microservice deployment with layer sharing in edge computing. IEEE Internet Things J. **11**(13), 23370–23383 (2024)
49. Shi, Y., Yi, C., Wang, R., Wu, Q., Chen, B., Cai, J.: Service migration or task rerouting: a two-timescale online resource optimization for MEC. IEEE Trans. Wirel. Commun. **23**(2), 1503–1519 (2024)
50. Tao, F., Zhang, H., Liu, A., Nee, A.Y.: Digital twin in industry: state-of-the-art. IEEE Trans Ind. Inform. **15**(4), 2405–2415 (2018)
51. Tao, Z., Xu, W., Huang, Y., Wang, X., You, X.: Wireless network digital twin for 6G: generative AI as a key enabler. IEEE Wirel. Commun. **31**(4), 24–31 (2024)
52. Van Huynh, D., Nguyen, V.D., Khosravirad, S.R., Sharma, V., Dobre, O.A., Shin, H., Duong, T.Q.: URLLC edge networks with joint optimal user association, task offloading and resource allocation: a digital twin approach. IEEE Trans. Commun. **70**(11), 7669–7682 (2022)
53. VanDerHorn, E., Mahadevan, S.: Digital twin: generalization, characterization and implementation. Decis. Support Syst. **145**, 113524 (2021)
54. Vrachnos, P., Ramonell, C., Koulalis, I., Ioannidis, K., Stipanovic, I., Vrochidis, S.: A framework for vision-based 3D inspections for maintenance activities and digital twin integration. In: Proceedings of IEEE CBMI, pp. 1–7. IEEE (2024)
55. Wang, D., Guo, J., Ouyang, Y., Wang, S., Yang, A., Ren, Z., Ding, Y., Chen, G., Zhou, C., Chen, D.: Leverage digital twins technology for network lifecycle management. In: 2022 IEEE 2nd International Conference on Digital Twins and Parallel Intelligence (DTPI), pp. 1–5. IEEE (2022)
56. Wang, Y., Feng, S., Zeng, Y., Sun, S., Tan, P.H.: Towards real-time digital twin of physical reality via intelligent wireless resource allocation. Mob. Comput, IEEE Trans (2025)
57. Wu, J., Chen, R., Yi, C.: A DRL-based deviation-aware federated digital twin construction over wireless edge network. In: International Conference on Wireless Artificial Intelligent Computing Systems and Applications (WASA), pp. 196–208. Springer (2025)
58. Wu, Y., Zhang, K., Zhang, Y.: Digital twin networks: a survey. IEEE Internet Things J. **8**(18), 13789–13804 (2021)
59. Xu, Y., Qamsane, Y., Puchala, S., Januszczak, A., Tilbury, D.M., Barton, K.: A data-driven approach toward a machine-and system-level performance monitoring digital twin for production lines. Comput. Ind. **157**, 104086 (2024)

60. Yang, H., Li, Y., Yao, K., Sun, T., Zhou, C.: A systematic network traffic emulation framework for digital twin network. In: 2021 IEEE 1st International Conference on Digital Twins and Parallel Intelligence (DTPI), pp. 94–97. IEEE (2021)
61. Yang, Y., Shi, Y., Chen, R., Yi, C., Kang, J.: Online optimization of edge empowered human digital twin deployment and task offloading. In: Proceedings of IEEE ICCC, pp. 856–861 (2024)
62. Yang, Y., Shi, Y., Yi, C., Cai, J., Kang, J., Niyato, D., Shen, X.: Dynamic human digital twin deployment at the edge for task execution: a two-timescale accuracy-aware online optimization. IEEE Trans. Mob. Comput. **23**(12), 12262–12279 (2024)
63. Yaqoob, M., Trestian, R., Tatipamula, M., Nguyen, H.X.: Digital-twin-driven end-to-end network slicing toward 6G. IEEE Internet Comput. **28**(2), 47–55 (2023)
64. Yu, Q., Ren, J., Fu, Y., Li, Y., Zhang, W.: Cybertwin: an origin of next generation network architecture. IEEE Wirel. Commun. **26**(6), 111–117 (2019)
65. Zafar, N., Ul Haq, I.: Traffic congestion prediction based on estimated time of arrival. PLoS ONE **15**(12), e0238200 (2020)
66. Zhang, Q., Yang, Y., Yi, C., Okegbile, S.D., Cai, J.: Energy- and cost-aware offloading of dependent tasks with edge-cloud collaboration for human digital twin. IEEE Internet Things J. **11**(17), 29116–29131 (2024)
67. Zhou, H., Chen, R., Yi, C., Li, J., Cai, J.: A three-party repeated coalition formation game for PLS in wireless communications with IRSS. In: Proceedings of IEEE WCNC, pp. 1–6 (2024)
68. Zhou, H., Chen, R., Yi, C., Zhang, J., Kang, J., Cai, J., Guizani, M.: A repeated coalition formation game for physical layer security aware wireless communications with third-party intelligent reflecting surfaces. IEEE Trans. Wireless Commun. (2025)
69. Zhu, Y., Chen, D., Zhou, C., Lu, L., Duan, X.: A knowledge graph based construction method for digital twin network. In: 2021 IEEE 1st International Conference on Digital Twins and Parallel Intelligence (DTPI), pp. 362–365. IEEE (2021)

Chapter 2
Key Issues and Enabling Technologies for DT Over Wireless Networks

2.1 Design Requirements and Challenges for DT Over Wireless Networks

DT rely on a tight integration of physical assets, sensing devices, computation resources, and communication networks to maintain accurate, real-time representations of dynamic environments. The design of DT over wireless networks faces unique challenges due to the heterogeneous, high-volume, and time-sensitive nature of the data involved. Ensuring high-fidelity modeling, seamless bidirectional interaction, and reliable communication demands careful consideration of both hardware and software foundations, as well as advanced network protocols that support deterministic, low-latency, and synchronized operations. In this section, we systematically examine the key requirements and technical challenges that must be addressed to realize robust and scalable DT over wireless infrastructures.

2.1.1 High-Fidelity DT Modeling

DT modeling demands hardware and software foundations to enhance robust, efficient, agile and scalable DT model construction, providing DT model with high-fidelity in long-term evolutions, i.e., adaptively and accurately synchronizing the state of the physical entities. The design requirements and challenges for DT modeling are illustrated in Fig. 2.1. We introduce the key design requirements and challenges for DT modeling from the perspectives of hardware and software foundations, as follows:

(1) *Hardware Foundations*

Hardware serves as the foundation of DT modeling, supporting accurate data acquisition, reliable communication, and efficient processing at the edge. Key design

C. Yi et al., *Self-Evolving Digital Twin Over Wireless Networks*, Wireless Networks,
https://doi.org/10.1007/978-3-032-08120-9_2

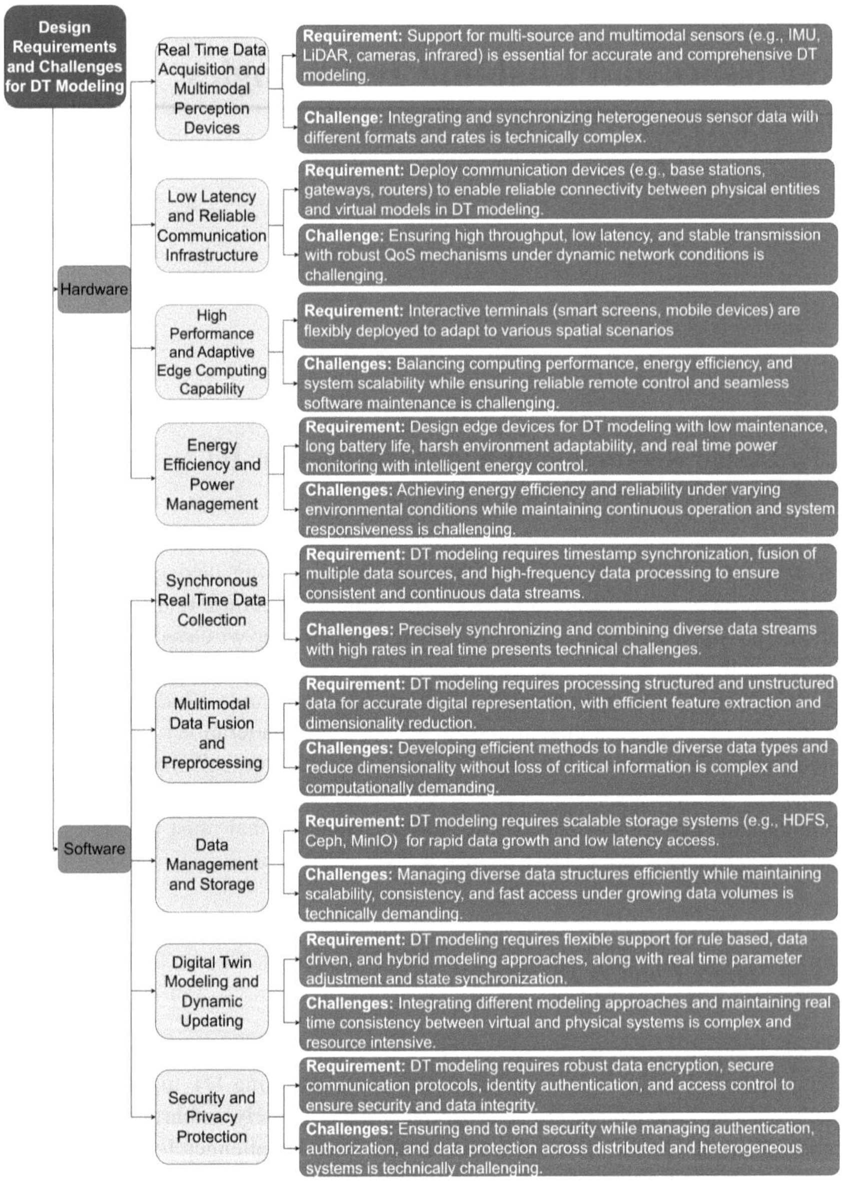

Fig. 2.1 Design requirements and challenges for DT modeling

requirements include real time multimodal sensing, low latency and dependable communication, localized edge computing for quick response, and power management with high energy efficiency. The following section presents the core technological requirements and corresponding challenges involved at the hardware layer.

- **Real Time Data Acquisition and Multimodal Perception Devices**: In DT modeling, real time data acquisition and multimodal perception devices form the foundation of the data layer, enabling the capture and transmission of dynamic and heterogeneous physical processes [4]. These devices influence the timeliness, completeness, and accuracy of the DT, directly affecting DT modeling and updating. To ensure temporal consistency, synchronized acquisition, edge preprocessing, timestamping, and buffering mechanisms are necessary. Sensors such as LiDAR, IMU, RGB-D cameras, GNSS, and environmental probes generate data in various formats and rates. Thus, hardware interoperability and unified data protocols must be implemented to support a consistent data ingestion pipeline [26, 154]. These requirements ensure system stability and sustained DT modeling.
- **Low Latency and Reliable Communication Infrastructure**: Building a low latency and reliable communication infrastructure is essential for DT modeling, ensuring continuous data transmission across heterogeneous environments [132]. Core components such as base stations, gateways, and routers must support stable communication with minimal jitter and packet loss. To support DT modeling and interaction, these systems must enable adaptive routing, edge preprocessing, and Quality of Service (QoS) prioritization [160]. Edge intelligence helps reduce communication delay by performing localized decisions. Redundancy and fault tolerance mechanisms must be integrated to ensure communication resilience. Meticulous hardware deployment and adaptive network control are needed to support scalable and uninterrupted DT updating [55].
- **High Performance and Adaptive Edge Computing Capability**: Edge computing enables DT systems to conduct localized processing and AI inference, reducing delays and supporting efficient DT modeling [153, 168]. This requires computing nodes with multicore processors, AI accelerators, sufficient memory, and stable operation under limited resources. Key capabilities include secure management, firmware updates, and orchestration across heterogeneous platforms. Challenges include synchronization latency, system scalability, and consistent performance under constrained conditions [34, 160]. Seamless integration with the DT communication backbone ensures accurate model synchronization and continuous DT updating, making edge computing essential for real time responsiveness and scalable DT deployment. Computational resource configuration and workload management at the edge require adaptive strategies to handle heterogeneous demands efficiently [91, 151].
- **Energy Efficiency and Power Management**: Energy efficiency is critical for edge devices used in DT modeling, especially under limited power budgets [44]. Devices performing continuous sensing and computation must use low power components while maintaining responsiveness. Hardware design should include dynamic voltage scaling, intelligent sleep scheduling, and energy aware task allocation [58]. Power management units and harvesting modules help extend device operation. Real time power monitoring and predictive control ensure system stability. Balancing processing needs with energy constraints supports sustained DT modeling and updating. Robust design improves long term reliability and enables scalable DT modeling across resource constrained settings.

(2) *Software Foundations*

Software plays a critical role in DT modeling by enabling seamless data collection, fusion, management, modeling, and security. Essential design requirements include synchronous real time data acquisition, effective multimodal data fusion and preprocessing, scalable data storage and management, accurate DT modeling and dynamic updating, as well as robust security and privacy protection. What follows is an overview of the key technological demands and challenges within the software domain.

- **Synchronous Real Time Data Collection**: Synchronous real time data collection is essential for DT modeling, ensuring temporally aligned data streams from heterogeneous sources such as sensors, IoT devices, and control systems [104]. Accurate DT modeling and updating require high temporal consistency to support simulation, diagnostics, and control. This depends on timestamping, clock synchronization via PTP or NTP, buffering strategies, and messaging protocols like MQTT, Kafka, and ROS 2. Middleware platforms must support concurrent data ingestion while preserving a unified timeline [145]. Challenges include asynchronous data sources, variable latency, and increasing sensor volume [173]. Ensuring precise synchronization at the data layer is key to maintaining system responsiveness and model fidelity during DT modeling.
- **Multimodal Data Fusion and Preprocessing**: In DT modeling, multimodal data fusion integrates various inputs such as imagery, point clouds, inertial signals, and environmental data [63]. Effective preprocessing, including noise filtering, normalization, and outlier removal, is required to improve data quality prior to fusion. Fusion techniques must support alignment at feature and decision levels and address uncertainty. Middleware must enable scalable, real time processing of diverse data streams with minimal delay [17, 93]. Key challenges include format heterogeneity, temporal misalignment, and missing inputs. Efficient fusion improves the reliability of DT modeling and supports consistent system updating under dynamic conditions.
- **Data Management and Storage**: Data management and storage support DT modeling by organizing and accessing large scale, heterogeneous data from sensors, edge nodes, and platforms [85]. The architecture must support flexible schemas, metadata integration, and semantic interoperability. Distributed storage systems like HDFS, Ceph, and MinIO offer scalability, fault tolerance, and responsive access for DT updating. Data frameworks should manage version control, provenance, and tiered storage of raw, processed, and archived content. Ensuring consistency, security, and compliance are key challenges [22, 107]. Robust storage infrastructure enables efficient DT modeling and continuous operation within distributed environments.
- **Digital Twin Modeling and Dynamic Updating**: DT modeling involves high fidelity modeling and continuous dynamic updating. Models must represent geometric structures, dynamics, and behaviors, using sensor data, environmental inputs, and operational logs. Modular design allows subsystems to update independently for scalable DT modeling [19]. Dynamic updating synchronizes the

DT with its physical entity by ingesting multi source data, calibrating models, and compensating for drift [82, 135]. Intelligent algorithms support anomaly detection and operational correction. Capabilities such as historical tracking, rollback, and version control ensure accuracy. Balancing complexity, uncertainty, and responsiveness is essential to sustained DT modeling.

- **Security and Privacy Protection**: Security and privacy are critical in DT modeling due to the sensitivity of data collected, transmitted, and stored. Key strategies include identity authentication, access control, and encryption of data in transit and at rest [3, 80]. Techniques like anonymization, differential privacy, and secure multi party computation protect data while enabling collaboration. Fine grained auditing and access control within the software architecture ensure compliance. As DT modeling scales and cyber threats evolve, security mechanisms must support sustainable and trustworthy DT updating. Ensuring data integrity and confidentiality is foundational for all DT modeling stages.

2.1.2 Underlying Network and Communication Foundation for DT

2.1.2.1 Bidirectional Physical-Virtual Interaction

Bidirectional physical–virtual DTinteraction establishes a dynamic bridge between the real and digital domains, enabling users to engage with DT models in an immersive and responsive manner [156]. Unlike conventional interaction paradigms, it emphasizes continuous two-way exchanges, where physical actions and contextual data are mapped into the virtual space, and virtual feedback is projected back into the physical environment in real time. This process relies on advanced multimedia devices such as Augmented Reality (AR) and Virtual Reality (VR) headsets, spatial audio systems, and haptic feedback tools, supported by networked DT infrastructures that ensure low-latency responsiveness and high reliability. The resulting interaction framework highlights immersive experience, adaptive environments, and collaborative engagement, while posing unique challenges at both the user-side hardware layer and the network-oriented software layer [161]. To address these challenges, the following sections examine the requirements for realizing effective, scalable, and trustworthy bidirectional DT service interaction. The overall content is shown in Fig. 2.2.

Hardware: In the process of DT service interaction, hardware devices not only assume the basic functions of data acquisition and output, but also realize the key carrier of virtual and real integration and immersive interaction. Hardware level design requirements and challenges are mainly from the following aspects.

- **Immersive User Devices for Bidirectional Interaction**: To create immersive DT service interactions, diverse devices are required, such as AR / VR headsets for 3D

	Technical Points	Core Technologies	Function
Seamless User-DT Interaction Technologies	Integrated Multimedia Interaction Platforms	AR/VR headsets; Ambisonics; Microsoft Surface Hub	Seamlessly blend physical and virtual environments via immersive displays, spatial audio, and multi-touch interfaces for rich interactive experiences.
	AI-Powered Sensor Fusion Modules	NVIDIA JETSON Sensor fusion chip; OpenPose; Fusion algorithm	Fuse multisource sensor data to generate accurate real-world representations and enable real-time, context-aware interactions
	Universal Device Integration Frameworks	Zigbee; OpenHAB; WebBluetooth	Integrate diverse hardware devices to enable seamless communication and unified interaction across the DT environment.
	Real-Time Multimedia Event Processing Engines	Apache Kafka; Apache Flink; Unity VR	Process and route high-volume, low-latency data in real time to enable dynamic and responsive interactions within DT environments.
	Intelligent Context-Aware Service Orchestration	Google Awareness; Camunda BPMN; IFTTT	Orchestrate service workflows dynamically based on real-time context to enable adaptive and user-specific DT interactions.
	Real-Time Collaborative Interaction Platforms	Firebase; WebRTC; Socket.io	Enable synchronized multi-user interactions to support collaborative decision-making, design, and monitoring across DT environments.

Fig. 2.2 Design requirements and challenges for bidirectional physical-virtual interaction

visualization, holographic projectors for embedding virtual objects in real spaces, spatial audio systems for realistic sound positioning, and haptic feedback tools for touch sensations [53]. These devices should work together to deliver seamless, high-fidelity multisensory feedback, making users feel truly present in both the real and virtual environments. The main challenge is ensuring synchronization across all sensory modalities. Any delay or mismatch between visual, auditory, or tactile signals can break immersion [134]. Devices must be both comfortable for long-term use and adaptable to different environments, requiring robust hardware integration and real-time coordination [51].

- **High-Precision User Sensing and Tracking for Physical-to-Virtual Mapping**: Accurate tracking is essential for DT service interaction. Cameras, depth sensors, and wearables capture user gestures, postures, facial expressions, and location,

translating these inputs into DT model manipulations. Multiple users may interact simultaneously, requiring the system to distinguish and synchronize individual inputs reliably [89]. Challenges include fusing data from various sensors, coping with environmental interference (such as poor lighting or occlusion) and maintaining stable performance in complex or crowded spaces. Real-time data transmission and robust tracking algorithms are crucial for smooth, responsive interaction [103].
- **Adaptive Deployment of Edge Interaction Terminals**: Interaction terminals such as projectors, smart displays, and mobile devices should be deployed flexibly throughout the environment, adapting to the movement and context of the user [94]. This ensures that users can access DT services from anywhere, with terminals automatically adjusting their position and output to suit the environment. The challenge is to enable terminals to self-configure, communicate seamlessly, and maintain uninterrupted data flow, even as users move or switch devices. Space awareness, dynamic resource allocation, and automated configuration are key technical hurdles [56].

Software: The software level is the "brain" of the DT service interaction, which is responsible for the management of multimodal inputs, the orchestration of interaction logic, the adaptation of user interfaces, and the support of multi-user collaboration. Its design requirements and challenges will be carried out from the following aspects.

- **Adaptive User Interfaces for Bidirectional Feedback**: The interaction of the DT service thrives on interfaces that adapt to the roles, preferences, and devices of the users. The system must generate personalized multimedia user interfaces, whether detailed manipulation controls for engineers or high-level dashboards for managers, while maintaining consistency across devices [62]. Developing adaptive User Interface (UI) frameworks and user profile algorithms is challenging, requiring real-time context recognition and seamless interface transitions as users move between terminals or environments [62].
- **Scalable Multi-User Collaboration over DT Networks**: DT service interaction increasingly involves multiple users working together on the same DT model. The model must enable real-time collaboration, including permission management, state synchronization, and conflict resolution. [66] Challenges include managing concurrent edits, synchronizing data between all users, and ensuring scalable performance as collaboration grows. Distributed frameworks and synchronization protocols are required to support industrial teamwork and virtual social spaces [15].

2.1.2.2 Advanced Wireless Communication for Supporting DT

The realization of robust and high-fidelity DT is critically dependent on the underlying wireless communication infrastructure. Unlike conventional networks optimized for best-effort data delivery, DT require deterministic, resilient, and highly synchronized wireless links to ensure the accuracy, responsiveness, and reliability of the virtual representation. Achieving this necessitates overcoming significant technical hurdles across multiple dimensions of wireless communication. The core challenges include ensuring continuous and lossless ingestion of heterogeneous sensor data streams under dynamic conditions, achieving ultra-precise time synchronization across distributed and resource-constrained devices, maintaining graceful degradation and critical data prioritization under spectrum scarcity, guaranteeing deterministic performance for interactive user commands and high-fidelity multimodal feedback, enabling seamless mobility across diverse and disjoint network domains, and providing robust privacy protection without violating stringent latency constraints. Addressing these interconnected challenges is paramount for realizing the full potential of DTs in complex industrial environments.The overall content is shown in Fig. 2.3.

- **Continuous and Lossless Data Ingestion**: The user plane of the radio interface should maintain a low packet loss rate, and in the absence of operator intervention, each construction domain should maintain a sufficiently high instantaneous uplink rate [133]. In DT construction, physical sensing networks exhibit heterogeneous data generation patterns. When a physical asset is first mirrored, thousands of heterogeneous sensing elements begin to emit streams that differ in rate, granularity and urgency. Visual sensors may produce dense point clouds at hundreds of megabits per second while temperature probes emit a few bytes every second. All of these streams must arrive at the model construction engine in their original temporal order and with no missing samples, because even a single dropped frame can corrupt the calibration of a physics based solver or mislead an anomaly detection routine [139]. The wireless medium, however, is governed by fading, shadowing and interference events that arrive without warning. Maintaining absolute integrity across such a stochastic channel demands coding strategies, retransmission policies and diversity schemes that do not yet exist in commodity radios. Furthermore, the instantaneous aggregate rate can surge by orders of magnitude whenever the asset undergoes a transient event such as a turbine stall or a structural vibration test. The network must therefore possess an elastic capacity that can expand within milliseconds and then contract just as quickly, all while respecting strict bounds on latency and power. Meeting this requirement remains an open challenge, because current medium access schemes are optimised for average rather than peak behaviour and because spectrum regulators have not yet defined a licensing framework that allows such rapid, on demand expansion [2].
- **Synchronised Sampling and Timestamp Preservation**: Every measurement must be stamped with a globally traceable time reference. The construction of DT relies on the fusion of multi-sensor measurement values for precise model calibration, so the reliability of DTs depends on the synchronization of sampling from their data sources [81]. When vibration data from a gearbox and thermal data from its bearing are fused, any misalignment greater than a few milliseconds will create phantom phase shifts that propagate into erroneous lifetime predictions. Achieving such tight synchronisation over wireless links is complicated by the fact that each sensor operates on an independent local clock whose drift is driven by temperature gradients and ageing crystals. Existing clock synchronisation pro-

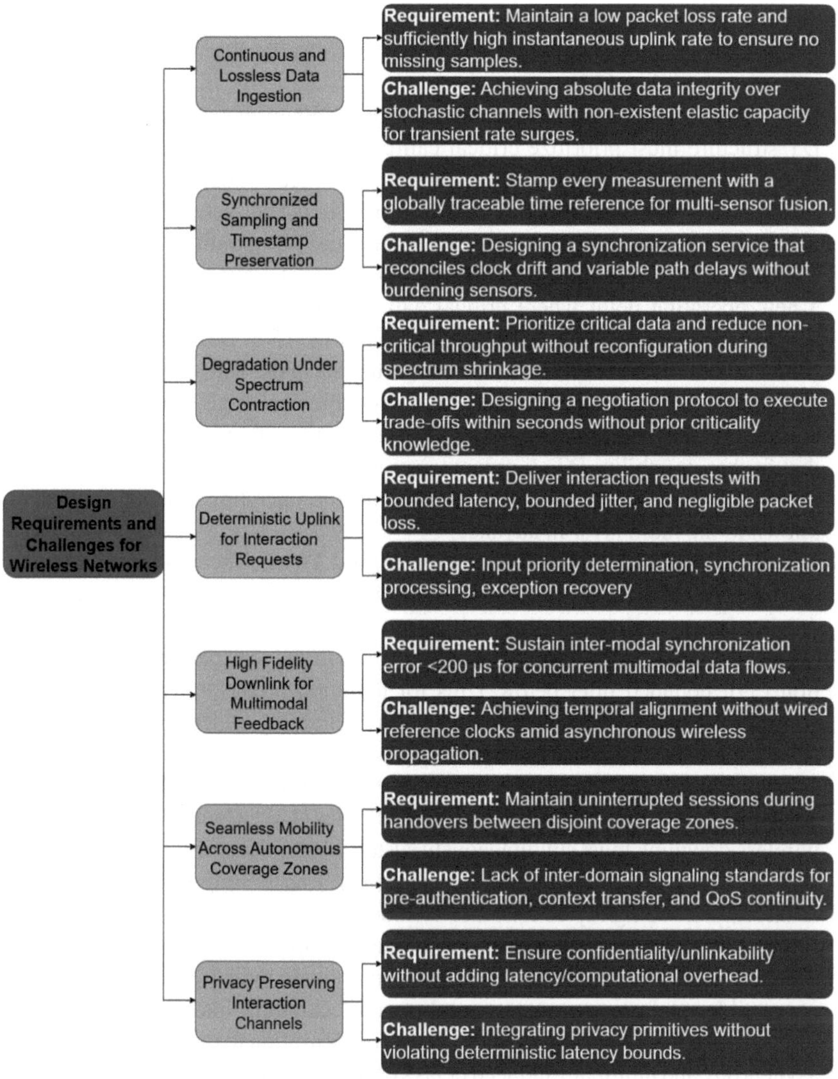

Fig. 2.3 Design requirements and challenges for advanced wireless communication for supporting DT

tocols such as Precision Time Protocol(PTP) or Network Time Protocol(NTP) were designed for wired backbones and assume frequent, symmetric exchanges that are impossible in duty cycled sensor networks [92]. Wireless propagation adds variable path delays, while energy harvesting introduces unpredictable sleep cycles. Additionally, multi-hop relaying introduces cumulative timing errors that exceed the bound unless each hop compensates with sub-microsecond precision, a capability absent in commercial low-power radios. Therefore, the challenge is to design a synchronisation service that can reconcile all of these uncertainties without burdening the sensors with additional radio traffic that would shorten their operational lifetime [57].

- **Degradation Under Spectrum Constriction**: When the spectrum shrinks, the DT structure must prioritize critical calibration data and reduce the throughput of non-critical data without reconfiguration messages. Industrial regulators may reallocate portions of the radio spectrum, neighbouring networks may generate unexpected interference, or a sudden surge in coexisting traffic might saturate the channel. In such moments the link must not collapse; instead, it must negotiate a new operating point where critical data continues to flow while non critical data is deprioritised or temporally decimated [142, 150]. Designing a negotiation protocol that can execute this trade off within seconds, without prior knowledge of which data subsets are critical, is a formidable challenge. Current congestion control mechanisms are reactive and coarse grained, often leading to oscillations that further destabilise DT. A proactive, model driven approach is needed, yet the parameters that define criticality evolve with the operating context and are not known to the radio layer in advance.
- **Deterministic Wireless Interaction for Digital Twins**: The wireless uplink must deliver time-critical interaction requests originating from end-user devices (e.g., AR headsets, haptic controllers, tablets) to the Digital Twin with stringent guarantees: bounded latency not exceeding 10 ms to prevent performance degradation [39], bounded jitter below 1 ms for temporal predictability in closed-loop control, and negligible packet loss to ensure reliability, independent of instantaneous channel state and despite mobility or variable load. These requests, typically small (tens of bytes to a few kilobytes), demand novel protocols surpassing the statistical guarantees of current radio technologies [143]. Complementing this, the wireless downlink must simultaneously sustain high-fidelity multimodal feedback at aggregate rates up to 5 Gbps per user [75], transmitting concurrent streams of uncompressed/lightly compressed stereoscopic video, spatial audio, and haptic force vectors [129]. Crucially, this downlink requires maintaining an extremely low inter-modal synchronization error within a single radio frame to prevent user-perceptible artefacts, necessitating advanced techniques like cross-modal framing, adaptive buffering, and predictive pre-compensation to overcome the inherent asynchrony and multipath-induced delays of wireless channels without a wired reference clock. Achieving this dual requirements of deterministic uplink command passing and synchronization, as well as high-throughput multimodal downlink, is the foundation of reliable and immersive DT interaction.
- **Seamless Mobility Across Autonomous Coverage Zones**: Interactive sessions must remain uninterrupted as users move across disjoint coverage islands created by private industrial networks. Industrial DT interactions involve handovers between 5G private/macro networks [9]. In industrial campuses, users frequently move between indoor private 5G networks, outdoor macro cellular networks, and temporary ad hoc networks deployed for maintenance or emergency scenarios. Each network may enforce distinct authentication protocols, security policies, and QoS frameworks. Handover procedures must preserve the state of the interactive

session including cryptographic keys, buffer occupancies and synchronisation offsets. The challenge is compounded by the fact that latency budgets for interaction traffic are stricter than those for conventional mobile data services. The primary challenge is the lack of inter-domain signalling standards for pre-authentication, context transfer, and QoS continuity. Existing mobility management protocols assume homogeneous access technologies and cannot handle rapid transitions between licensed and unlicensed bands. Security handshakes add latency that can exceed the interaction deadline. Additionally, state synchronisation of the DT across domains requires low-latency backhaul that may not be available in private or temporary networks.

- **Privacy Preserving Interaction Channels**: The wireless link must ensure confidentiality and unlinkability of interaction traffic without introducing additional latency or computational overhead that violates interaction deadlines [6]. DT service interactions traffic reveals sensitive contextual information such as user identity, behavioural patterns, and proprietary operational data [80]. Encryption alone is insufficient because traffic volume, timing, and directional patterns can be exploited for inference attacks. The requirement is to provide differential privacy guarantees at the link layer while maintaining deterministic latency. Challenges include the computational cost of privacy-preserving primitives such as differential privacy noise injection or onion encryption, which can violate latency bounds. Traffic shaping to obfuscate patterns increases buffer occupancy and jitter. Cross-layer Optimization is required to integrate privacy mechanisms with radio resource allocation, but current protocol stacks lack standardised interfaces for such coordination.

2.2 Enabling Technologies for DT Over Wireless Networks

In this section, we discuss the enabling technologies for realizing high-fidelity DTs and maintaining real-time synchronization with their physical counterparts. Along the life cycle of a DT model or service, these technologies are categorized into: i) multimodal data collection and fusion, which provide the fundamental dataset for DT model training; ii) accurate and comprehensive DT modeling, which yields models with strong replication capability; iii) seamless user–DT interaction, which enables customized DT services tailored to user requirements; and iv) evolution-aware wireless communication, which ensures efficient and timely transmission of both data and models to mitigate the adverse effects of excessive latency.

(1) *Multimodal Data Collection and Fusion Technologies*

Multimodal data collection and fusion serve as the foundation for reliable and dynamic DT modeling. Real-time data acquisition and perception devices enable the capture of diverse data types through high-performance sensors, ensuring comprehensive environmental awareness. Synchronous real-time data collection guarantees temporal and spatial consistency of distributed sensor inputs, supported by precise synchronization and low-latency communication. Multimodal data fusion and preprocessing further enhance data quality by aligning, cleaning, and integrating heterogeneous sources for robust model construction. Efficient data management and storage frameworks sustain large-scale, high-velocity data flows, providing secure and scalable support for DT applications. Finally, energy efficiency and power management technologies ensure sustainable operation of distributed sensing and computing devices, balancing real-time responsiveness with resource constraints. Together, these enabling technologies establish a coherent and resilient foundation for DT modeling.

Enabling Technologies for Real Time Data Acquisition and Multimodal Perception Devices: Real time data acquisition and multimodal perception devices, as key enablers of DT modeling, integrate high performance sensors to capture and process diverse data types such as point clouds, imagery, audio, environmental metrics, and positioning information. The following provides a systematic overview of core enabling technologies and their hardware functions.

- **LiDAR**: LiDAR technology generates high resolution 3D point clouds by emitting laser pulses and measuring the time of flight of reflected signals. This enables precise distance measurement critical for constructing detailed spatial and topological models of complex physical environments. Multi beam LiDAR units such as 16, 32, and 64 line variants are extensively utilized in DT modeling scenarios, delivering rich environmental data across multiple vertical layers [81, 159]. These sensors support applications including high fidelity building reconstruction, dynamic obstacle detection in urban and industrial settings, and real time localization and mapping for autonomous agents such as UAVs, ground vehicles, and autonomous cars. Qian et al. [89] proposed a physically accurate LiDAR simulation method for automotive digital twins. Their approach integrates ray tracing with a real world infrared reflectivity database, enabling realistic modeling of both geometric and signal properties.
- **IMUs**: IMUs provide continuous measurements of linear acceleration and angular velocity across three axes, serving as essential components for motion tracking, orientation estimation, and dynamic state sensing. Typically integrating gyroscopes, accelerometers, and magnetometers, these compact and power efficient devices capture rapidly changing motion dynamics at high sampling rates between 100 and 1000 Hz [63, 105]. IMUs are crucial for DT modeling involving mobile platforms such as robots, autonomous vehicles, drones, and industrial machinery, enabling real time synchronization of position, velocity, and orientation between physical entities and their digital counterparts. Their integration with edge computing nodes facilitates local state estimation, thereby reducing communication latency and easing central processing loads [79, 81].
- **GNSS**: GNSS modules deliver precise absolute positioning by receiving satellite signals from constellations including GPS, GLONASS, Galileo, and BeiDou. High precision RTK GNSS units enhance localization accuracy to the centimeter level through differential corrections from reference stations. This precision is indispensable for applications such as precision agriculture, autonomous navigation,

traffic simulation, and large scale smart city mapping, where exact spatial alignment between physical environments and digital models is essential for reliable DT modeling [72]. GNSS also provides accurate timing references, enabling synchronization across distributed DT components, supporting coherent multisensor fusion, consistent data logging, and coordinated real time control [65].

- **Infrared and Thermal Imaging Sensors**: Infrared and thermal sensors enable non invasive monitoring of heat distribution and thermal anomalies in real time. These sensors are instrumental in continuous thermal assessment applications, including smart building management, industrial equipment diagnostics, and critical infrastructure monitoring [111, 162]. Thermal data contributes essential inputs for modeling heat flow, thermal stresses, and potential hazard zones within DT modeling processes, thereby facilitating predictive maintenance, HVAC optimization, and compliance with safety standards.
- **Microphone Arrays**: Microphone arrays are fundamental for sound aware DT modeling, enabling real time acoustic perception and interaction. By capturing spatial temporal audio features, these arrays allow digital models to replicate auditory environments with high fidelity. Advanced beam forming and noise suppression techniques facilitate precise localization of sound sources and signal isolation even in noisy or reverberant conditions [26, 31]. Applications include voice driven smart building controls, occupancy detection, acoustic anomaly monitoring, and security oriented audio surveillance such as detecting breaking glass, distress calls, or machinery malfunctions. Integrating audio perception enhances immersive, context aware, and human centric DT modeling.

Enabling Technologies for Synchronous Real Time Data Collection: Synchronous real time data collection is fundamental in DT modeling to ensure timely and coherent integration of sensor data across distributed systems. The following outlines key enabling technologies supporting this capability.

- **Time Synchronization Protocols**: Two main protocols used in DT modeling are the Precision Time Protocol (PTP) and the Network Time Protocol (NTP). PTP provides sub microsecond synchronization precision over Ethernet by exchanging timing messages between master and slave clocks, dynamically correcting offsets and drifts. Its high accuracy is essential in industrial grade DT modeling requiring precise time alignment for deterministic correlation of distributed sensor data [40, 158]. NTP offers millisecond level accuracy and prioritizes robustness and scalability over IP networks. While less precise than PTP, NTP is suited for DT modeling over wide area or cloud networks with variable latency and jitter [61]. Both protocols play critical roles in ensuring temporal consistency in DT data integration.
- **Middleware and Messaging Systems**: Middleware solutions such as Data Distribution Service (DDS), Apache Kafka, and MQTT enable efficient data exchange in DT modeling. DDS is a real time publish subscribe protocol designed for high performance, reliability, and scalability. It offers fine grained quality of service controls to manage latency, ordering, and durability, crucial for DT modeling that

demands strict timestamp sequencing and guaranteed delivery of sensor streams. DDS supports decentralized peer to peer communication, enhancing fault tolerance [19]. Apache Kafka is a distributed streaming platform that manages large heterogeneous sensor data streams with high throughput and fault tolerance. Kafka's architecture decouples data producers and consumers, ensuring scalable, ordered data retention suitable for real time analytics in DT modeling [62]. MQTT is a lightweight publish subscribe protocol optimized for constrained devices and unreliable networks. It balances delivery reliability and latency with multiple quality of service levels, supporting continuous data streaming amid intermittent connectivity in DT modeling [130].
- **Data Buffering and Alignment**: Buffering and alignment manage asynchronous, heterogeneous sensor streams to ensure data coherence within DT modeling pipelines. Sliding window buffers temporarily store incoming packets, grouping and smoothing data based on timestamps to reduce effects of network jitter, packet loss, and sensor sampling inconsistencies. Time series databases like InfluxDB and TimescaleDB are key for storing and managing buffered data, providing precise timestamp indexing and efficient querying. These databases enable real time alignment and synchronization of multisource data, maintaining unified, time consistent datasets essential for accurate multimodal fusion and comprehensive downstream DT modeling and updating [31, 38]. Together, these techniques ensure that DT systems accommodate diverse inputs while preserving temporal integrity for reliable simulation and decision support.

Enabling Technologies for Multimodal Data Fusion and Preprocessing: Multimodal data fusion and preprocessing are critical in DT modeling to ensure heterogeneous sensor inputs are cleaned, aligned, and prepared for accurate integration. Key enabling technologies are summarized below.

- **Multimodal Data Fusion**: Fusion methods are categorized into data level, feature level, and decision level, each integrating multisource information differently. Data level fusion combines raw or preprocessed data directly from multiple sensors. Techniques such as weighted averaging, Kalman filtering–including Extended Kalman Filter (EKF) and Unscented Kalman Filter (UKF)–and Bayesian inference handle uncertainty and noise effectively, providing precise state estimation needed for environment modeling in DTs [89]. Feature level fusion integrates extracted features from different modalities into a joint feature space. Machine learning models like Convolutional Neural Networks (CNNs), Recurrent Neural Networks (RNNs), and Transformers learn complex cross modal correlations. This fusion improves computational efficiency and robustness, essential for scalable DT modeling. Decision level fusion merges outputs or confidence scores from independent models analyzing different modalities, using methods such as majority voting, Dempster Shafer theory, or ensemble learning [63]. It enhances robustness by compensating sensor failures and managing uncertainty during DT updating.
- **Data Preprocessing**: Preprocessing prepares raw sensor data for multimodal fusion by improving quality, correcting inconsistencies, and aligning data spa-

tially and temporally. Noise reduction enhances signal fidelity; image denoising uses median filtering, Gaussian smoothing, and bilateral filtering, while 3D point clouds apply statistical outlier removal and voxel grid filtering to remove noise and downsample data. Calibration corrects systematic measurement errors. Intrinsic calibration addresses sensor specific distortions such as lens aberrations and biases, while extrinsic calibration aligns coordinate frames among sensors to ensure spatial consistency. Accurate calibration is fundamental for reliable multimodal fusion in DT modeling. Feature extraction transforms raw data into descriptors for fusion and analysis. Techniques include keypoint detectors like SIFT, SURF, and ORB for images, and normal vector estimation and voxelization for point clouds. Other sensors provide statistical or domain specific features summarizing physical properties [10, 18]. These features support structured data representations that underpin downstream DT analytics and updating.

Enabling Technologies for Data Management and Storage: Efficient management of large scale, heterogeneous, and high velocity data is essential to ensure fidelity and timeliness of digital models during DT modeling. Key supporting technologies are summarized below.

- **Scalable Distributed Storage Systems**: Centralized storage often cannot meet the volume and diversity demands of DT modeling data. Distributed storage platforms such as HDFS, Ceph, Amazon S3, and MinIO provide elastic scalability, high availability, and fault tolerance, supporting reliable data persistence under heavy workloads. Object storage is suitable for unstructured sensor data like images, point clouds, and video. Time series databases including InfluxDB and TimescaleDB efficiently manage structured temporal data, enabling rapid querying, aggregation, and analytics. Together, these systems establish a robust infrastructure to handle heterogeneous DT data streams throughout construction and updating processes [33, 122].
- **Data Schemas and Metadata Management**: Flexible schema designs accommodate evolving data types and formats in DT modeling. Schema on read enables raw data storage with deferred parsing, improving adaptability. Unified metadata frameworks enhance data discoverability, provenance, and versioning, ensuring integrity and trustworthiness over the DT lifecycle. Semantic ontologies and linked data technologies enrich data context and promote interoperability across diverse DT components, supporting seamless integration and effective knowledge sharing during modeling and updating [42, 138].
- **Data Lifecycle and Version Control**: Managing dynamic data during DT modeling requires robust lifecycle and version control. Tools like Git based versioning, Delta Lake, and Apache Hudi efficiently track changes, supporting incremental updates, rollbacks, and reproducible simulations. These systems maintain detailed data lineage and audit trails, enhancing traceability and governance. Tiered storage architectures separate frequently accessed from archived data, optimizing cost and query efficiency. Automated lifecycle management enforces retention policies, archiving, and deletion to ensure compliance and sustainability [32, 81].

- **Real Time Querying and Analytics**: Real time querying and analytics are critical for timely decisions in DT modeling involving large heterogeneous data streams. Advanced databases such as NewSQL combine NoSQL scalability with traditional SQL transactional guarantees, supporting transactional and analytical workloads with low latency. In memory data grids improve performance by storing data in RAM for rapid parallel processing [46]. Time series databases like InfluxDB and TimescaleDB use compression and specialized indexing to accelerate queries for monitoring dynamic DT states during modeling and updates. Integration with big data and machine learning frameworks enables complex event processing, anomaly detection, and predictive maintenance, improving operational reliability during DT lifecycle management [73, 112].

Enabling Technologies for Energy Efficiency and Power Management: With widespread distributed sensing and computing in DT modeling, power efficiency is critical, especially for edge and mobile devices. Key enabling technologies focus on their roles and implementation in DT modeling.

- **Low Power Hardware Architectures**: Low power hardware designs are fundamental to enhancing energy efficiency in DT modeling, particularly for battery powered edge nodes, UAVs, and mobile sensors. These architectures minimize power consumption while maintaining performance for long term stable operation. ARM based processors are central, known for energy efficient performance per watt and broad embedded system adoption. Modern ARM Cortex A and Cortex M series chips offer heterogeneous multi core configurations and on chip low power modes, suitable for continuous DT data processing at the edge [144]. Complementing CPUs, specialized AI accelerators provide machine learning inference with minimal power use, enabling local analytics to reduce cloud dependence. Field Programmable Gate Arrays (FPGAs) offer adaptable, reconfigurable logic supporting energy efficient sensor fusion, control logic, and edge analytics. Unlike fixed units, FPGAs combine adaptability with power aware design, supporting responsive DT modeling and system behavior replication [87]. Passive cooling and thermal aware scheduling further reduce energy dissipation in embedded and outdoor settings.
- **Dynamic Voltage and Frequency Scaling and Power Gating Techniques**: Dynamic Voltage and Frequency Scaling (DVFS) and power gating enable fine grained energy management in power constrained devices such as edge nodes and embedded systems. These mechanisms adjust power consumption according to workload demands, extending battery life, reducing heat, and maintaining real time performance. DVFS modulates processor voltage and frequency based on computational load, lowering both during idle or light activity, and increasing them during peak processing [164]. Power gating reduces static power by switching off inactive functional blocks, eliminating leakage currents. Unlike DVFS tuning active components, power gating disables idle modules such as I/O controllers or accelerators, offering deeper energy savings. As semiconductor scales shrink, static leakage grows, making power gating essential in low power architectures sup-

porting continuous DT updating [20, 27]. Power Management Units or embedded firmware monitor system status and orchestrate DVFS and power gating, aligning consumption with workload fluctuations.

- **Renewable Energy Harvesting Devices**: Renewable energy harvesting devices support sustainable power for edge devices in DT modeling, especially where power availability is limited [108, 149]. Photovoltaic panels convert solar energy into electricity, commonly powering outdoor edge nodes and UAVs, with energy stored in batteries or supercapacitors. Wind energy harvesters capture kinetic energy from airflow at elevated locations. Vibration energy harvesters use piezoelectric or electromagnetic transducers to convert mechanical vibrations from infrastructure into electrical energy. Hybrid systems integrate multiple renewable sources to improve power reliability and maintain stable operation. These technologies contribute to autonomous sensor operation and reduce maintenance needs in DT deployments [120].
- **Power Management Integrated Circuits and Battery Technologies**: Power Management Integrated Circuits (PMICs) efficiently distribute power, managing multiple inputs and loads, stabilizing voltages, controlling charging, and switching among batteries, harvesters, and external sources. By optimizing power flow, PMICs reduce losses, enhance reliability, extend battery life, and ensure uninterrupted DT operation in constrained environments. Modern PMICs feature protections against overcharge, temperature extremes, and current surges, and support fast charging and balanced multi cell management [21]. Battery technologies such as lithium ion and lithium polymer dominate due to high energy density and cycle life. Emerging chemistries like solid state and lithium sulfur promise higher capacity, safety, and thermal stability, extending edge device operation. Integrating advanced PMICs with battery solutions supports sustainable DT modeling, facilitating extended operation and real time data acquisition critical for accurate DT modeling and dynamic updating [14].

(2) *Accurate and Comprehensive DT Modeling Technologies*

Accurate and comprehensive DT modeling relies on an integrated set of enabling technologies. Low-latency and reliable communication infrastructures provide the foundation for efficient data transmission and seamless interaction between physical and digital spaces. High-performance and adaptive edge computing enhances local data processing, reduces response times, and supports real-time decision-making. Advanced modeling and dynamic updating techniques ensure that DTs remain synchronized, high-fidelity, and responsive to changes in physical systems. Meanwhile, robust security and privacy protection technologies safeguard data integrity and confidentiality across the DT lifecycle. Collectively, these technologies establish the accuracy, adaptability, and trustworthiness required for comprehensive DT modeling.

Enabling Technologies for Low Latency and Reliable Communication Infrastructure: Efficient data transmission is essential for DT modeling, requiring communication modules that ensure low latency, high reliability, and adaptive connectivity. Key devices and strategies to enhance DT network performance include:

- **Industrial Grade Switches and Routers**: Industrial grade switches and routers form the communication backbone in DT modeling networks, providing robust, high speed routing across distributed components. Designed for harsh environments with temperature extremes, humidity, dust, and electromagnetic interference, they feature high throughput backplanes capable of managing gigabit traffic without degradation [136]. Equipped with redundant power supplies, dual uplink ports, and failover mechanisms, these devices guarantee uninterrupted data flow and high availability. Their ASIC based forwarding engines prioritize latency sensitive traffic such as sensor data, control commands, and actuator feedback, ensuring deterministic communication performance. EMI shielded metal enclosures, DIN rail mounting, and fanless heat dissipation further improve operational stability [118]. Such devices are vital to maintaining synchronized, deterministic, and low latency exchanges within DT modeling.
- **High Performance Antenna and RF Modules**: In complex environments, signal attenuation, multipath effects, and interference degrade communication quality. High performance antenna systems employ high gain directional antennas, multiple input multiple output (MIMO), and beam forming to optimize signal strength and coverage dynamically. RF front end modules integrate low noise amplifiers (LNAs), surface acoustic wave filters, and power amplifiers to improve signal clarity and robustness. LNAs amplify weak signals while maintaining low noise, and SAW filters suppress out of band interference, reducing error rates in noisy spectral conditions. These enhancements sustain reliable communication critical for DT modeling [37, 47, 86].
- **Redundant Ethernet Interface Cards and Network Failover Units**: To maintain continuous communication during DT modeling, redundant Ethernet hardware is deployed. Dual port Ethernet interface cards provide automatic link switching and seamless failover across independent network paths. Specialized redundancy protocols such as High availability Seamless Redundancy and Parallel Redundancy Protocol ensure zero packet loss under physical link failures. These cards are embedded in industrial edge gateways or automation controllers. Integrated modules supporting multi interface aggregation–combining wired Ethernet, WiFi, and cellular–enable dynamic path selection based on link quality. Hot swappable power modules and redundant power inputs further improve availability, minimizing downtime in critical DT infrastructure [28, 97].
- **Application Specific Integrated Circuit Chips for Traffic Prioritization and QoS Scheduling**: DT modeling involves concurrent data flows like control loops, telemetry feedback, and streaming data, requiring deterministic communication. ASIC chips embedded in industrial switches and routers provide built in quality of service scheduling engines that classify and prioritize traffic at the hardware level. They manage per port queues, implement strict priority and weighted fair queuing, and apply traffic shaping to delay tolerant streams. Dedicated memory buffers and high speed classifiers ensure latency sensitive packets transmit with minimal jitter and delay. By isolating critical traffic from background processes, these hardware accelerators maintain timing constraints essential for reliable DT synchronization and control [121, 128].

Enabling Technologies for High Performance and Adaptive Edge Computing Capability: Edge computing is essential for DT modeling by enabling low latency, real time interaction, and localized data processing. Key enabling technologies are summarized below.

- **Edge Storage**: Edge storage hardware is critical for efficient data management and reliable persistence in DT modeling. Operating in environments with constrained bandwidth or unstable networks, edge nodes use local caching and non volatile storage to reduce latency and avoid data loss. Core components include high speed NVMe SSDs, industrial grade modules such as eMMC and UFS, and cache units. NVMe SSDs provide fast read and write performance through PCIe interfaces, facilitating real time processing of large sensor datasets [70]. Industrial modules ensure robustness under challenging conditions, supporting stable data retention. Edge storage also enables distributed synchronization and backup across edge nodes, enhancing redundancy and fault tolerance. These capabilities underpin continuous state tracking and historical data availability, foundational for accurate simulation and analytics within DT modeling [13].
- **Edge AI Accelerators**: Edge AI accelerators are vital hardware for in situ processing and interpretation of sensor data in DT modeling. These devices provide dedicated computational resources for machine learning inference, computer vision, anomaly detection, and predictive analytics at the network edge [17]. Examples include NVIDIA Jetson, Google Coral TPUs, and Intel Movidius chips, optimized for high throughput neural network computations on raw inputs like images, point clouds, and time series data [5, 76]. Performing intensive analytics locally reduces latency and bandwidth usage, while enhancing data privacy by limiting raw data transmission. Edge AI accelerators support containerized deployment and flexible model updates, enabling scalable, modular DT infrastructures that adapt to evolving analytics requirements and sensor additions.

High Performance and Adaptive Edge Computing Capability				
Technical Points	**Core Technologies**			**Function**
Edge Storage	NVMe SSDs	eMMC UFS	Cache Modules	Enable efficient data management, low-latency access, and resilient local storage, ensuring data continuity and integrity for accurate and reliable DT modeling.
Edge AI Accelerators	NVIDIA Jetson	Intel Movidius Myriad	Google Coral TPU	Enable local AI acceleration for real time interpretation and anomaly detection, allowing Digital Twins to make timely decisions with reduced reliance on central computing.
Edge Gateway	Multi-protocol Interfaces	LoRa	Communication module	Integrate multiprotocol interfaces, multicore processors, and large memory to ensure seamless real time data processing at the edge.

Fig. 2.4 Enabling technologies for high performance and adaptive edge computing capability

- **Edge Gateway**: Edge gateways serve as essential hardware intermediaries bridging diverse sensor networks with centralized DT modeling platforms. Equipped with high performance multi core processors and large memory buffers, they manage massive real time data flows, ensuring efficient caching and transient storage under high throughput [152]. Supporting multiple communication protocols such as Ethernet, 5G, Wi Fi, CAN, RS 485, and LoRa, gateways integrate heterogeneous sensors into unified edge processing hubs. Gateways perform preprocessing including filtering, normalization, and precise timestamp synchronization on device, reducing raw data transmission volumes. This optimizes bandwidth usage and lowers latency, supporting responsive DT modeling and updating [48, 146]. Modern gateways enable containerized microservices, allowing flexible and scalable DT application deployment close to data sources [102, 106, 155] (Fig. 2.4).

Enabling Technologies for Digital Twin Modeling and Dynamic Updating: DT modeling and dynamic updating are essential to maintain accurate and synchronized representations of physical systems. High fidelity, timely responsiveness, and adaptability remain central in DT modeling. Key supporting technologies include:

- **Geometric Modeling and 3D Reconstruction**: Geometric modeling provides the spatial foundation of DT modeling, creating accurate three dimensional representations of assets and environments. Common methods include LiDAR scanning, photogrammetry, structured light scanning, and Building Information Modeling (BIM). Tools such as Autodesk ReCap and open source libraries like OpenMVG and OpenMVS facilitate these processes. CAD software including AutoCAD and SolidWorks ensures engineering precision, while Blender and 3ds Max support detailed mesh editing. These models enhance visual fidelity and serve as reference frameworks for overlaying sensor data and simulations [62, 81]. Advances in 3D reconstruction and point cloud processing improve the efficiency and accuracy of spatial models [17].
- **Physical and Behavioral Modeling**: Physical modeling enables DT modeling to simulate multiphysics phenomena including dynamics, thermal transfer, fluid mechanics, and electromagnetic fields. Standards like Modelica, FMI, and simulation tools like ANSYS Twin Builder support integration of heterogeneous behaviors [137]. Behavioral modeling captures logical and reactive dynamics using platforms such as Stateflow and AnyLogic for discrete event and agent based simulations. Languages like SysML and tools such as MagicDraw support system level validation. Formal methods including state machines and Petri nets represent concurrency and error handling. Combining physical, behavioral, and data driven models enhances DT modeling accuracy and flexibility [18].
- **Dynamic Simulation Engines**: Simulation engines such as Unity3D, Unreal Engine, and CesiumJS support DT modeling through 3D visualization and dynamic system interaction. These platforms render complex scenes, integrate data streams, simulate physical processes, and visualize analytics efficiently. Unity3D and Unreal Engine enable modular simulation of machinery and control systems using

Digital Twin Modeling and Dynamic Updating		
Technical Points	**Core Technologies**	**Function**
Geometric Modeling and 3D Reconstruction	Blender, AutoCAD, 3ds Max	Provide the foundational 3D spatial framework for DT construction, enabling accurate representation, simulation, and sensor data integration of physical assets.
Physical and Behavioral Modeling	Simulink, Gazebo, Modelica	Enable DTs to simulate multiphysics processes and control logic, offering predictive insights into system behavior under changing conditions.
Dynamic Simulation Engine	Unity3D, UE5, CesiumJS	Enable high fidelity modeling and photorealistic rendering to support immersive and interactive DT construction.

Fig. 2.5 Enabling technologies for digital twin modeling and dynamic updating

middleware like ROS and MQTT. CesiumJS focuses on large scale geospatial visualization for infrastructure level DT modeling [117]. These engines support DT updating by enabling user interaction and feedback, ensuring synchronized evolution of virtual and physical systems [23] (Fig. 2.5).

Enabling Technologies for Security and Privacy Protection: Security and privacy are fundamental for ensuring integrity and data protection throughout DT modeling, updating, and operation. Key enabling technologies include:

- **Data Encryption Technologies**: Encryption ensures confidentiality and integrity across the DT modeling lifecycle. Transport Layer Security (TLS) and Secure Sockets Layer (SSL) secure data in transit, protecting sensor streams, control messages, and system interactions within the DT infrastructure [35]. End to End Encryption (E2EE) encrypts data from source to destination without revealing plaintext, supporting secure telemetry transfer in DT modeling. Data at Rest Encryption uses symmetric methods like Advanced Encryption Standard (AES) to protect stored data from unauthorized access during DT lifecycle management [95]. Advanced cryptographic techniques such as Homomorphic Encryption and Secure Multi Party Computation (SMPC) enable computation on encrypted data without exposure, enhancing privacy in DT modeling and analysis.
- **Access Control and Identity Authentication**: Robust access control and authentication safeguard DT modeling platforms from unauthorized access. Role Based Access Control (RBAC) restricts permissions based on user roles such as operator, analyst, or administrator, reinforcing least privilege policies in DT modeling and updating. Attribute Based Access Control (ABAC) includes contextual variables like user location, device, and time, enabling dynamic policy enforcement [71]. Multi Factor Authentication (MFA) combines credentials and biometrics to strengthen user verification. Blockchain ensures tamper proof audit trails through decentralized ledgers, improving accountability and transparency in DT platforms [80, 125].

- **Privacy Protection Mechanisms**: Protection of personal and sensitive operational data is crucial during DT modeling. Differential Privacy injects calibrated noise during analysis to prevent re identification while preserving statistical utility in DT modeling. Data Anonymization and De Identification remove or mask attributes such as names, coordinates, and device IDs before processing, ensuring compliance in DT updating [83]. Federated Learning trains models locally on edge devices, with only parameter updates shared, preserving privacy under frameworks like GDPR and HIPAA [100]. Access Auditing and Compliance Monitoring enforce privacy policies through systematic logging and automated controls [110].

(3) *Seamless User-DT Interaction Technologies*

In the realm of DT service interactions, the bridge between the physical and virtual worlds is paramount. The essence of these interactions lies in providing immersive and intuitive experiences that allow users to interact with digital representations of physical assets or systems in real-time. To facilitate these dynamic exchanges, a robust set of enabling technologies plays a pivotal role. These technologies span across hardware and software systems, each improving different facets of the interaction process. In this section, we explore how these technologies underpin virtual-physical fusion and contribute to seamless, context-aware service orchestration in DT environments [84]. We focus particularly on how hardware devices enable immersive interaction and how software systems power real-time, adaptive, and multimodal responses during the DT service interaction [23]. An overview of all corresponding technologies is illustrated in Fig. 2.6.

Enabling Technologies for Integrated Multimedia Interaction Platforms: Integrated multimedia interaction platforms form the core of a DT interaction, enabling users to seamlessly blend and navigate between physical and virtual environments. These platforms involve a combination of immersive display systems, spatial sound technologies, and multi-touch interfaces to provide a rich and interactive experience [147].

- **AR/VR Headsets**: AR and VR headsets serve as essential hardware devices for creating an immersive DT service interaction. These headsets act as windows through which users experience and engage with DTs of real-world assets, environments, or systems. VR headsets like the Meta Quest allow users to be fully immersed in a virtual environment, while AR devices like the Apple Vision Pro enable the overlay of digital information onto the physical world. This combination of AR and VR creates a hybrid space where the virtual and physical worlds coexist, enhancing user engagement [17]. In a typical DT service interaction, an engineer could use a VR headset to interact with a DT of a factory floor, monitoring production line conditions, while simultaneously using AR to visualize data overlays on physical machinery in real-time. This fusion of AR and VR allows for highly dynamic and intuitive interactions with the DT model, where actions such as manipulation, visualization, and monitoring can be carried out through natural gestures, gaze, or voice commands [126].

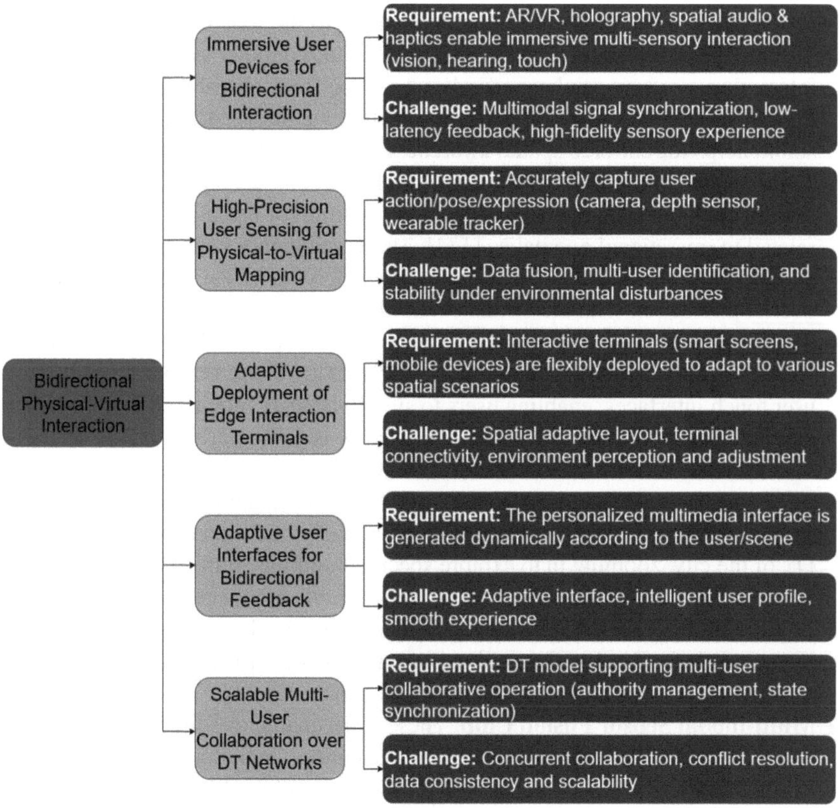

Fig. 2.6 Seamless user-DT interaction technologies

- **Holographic Projection Equipment**: Holographic projection systems, such as Holofan, provide a cutting-edge method for visualizing DTs in a 3D space without the need for wearables. These systems project 3D holograms into the air, offering users the ability to view and interact with digital models of physical objects in a fully immersive way [52]. This technology significantly enhances the virtual-physical interaction by allowing users to see a DT of a complex piece of equipment or a large infrastructure like a building model, all in 3D. In DT services, these holographic projections facilitate collaborative design, maintenance, or inspection processes. For instance, a team of engineers could collaboratively interact with a holographic 3D model of a DT of a building to assess the structural integrity or energy efficiency, while physically being in the same room [127]. The ability to freely walk around and manipulate the hologram in 3D space makes it easier to analyze complex data and make informed decisions quickly.
- **Space Sound Systems**: Spatial audio, powered by technologies like Ambisonic Loudspeakers, plays a crucial role in creating an immersive multi-sensory experience in DT service interactions. By integrating spatial sound systems, users are

able to experience audio cues that correspond to the location and movement of virtual objects within a DT environment. This adds another layer of realism to the interaction, as users can hear sounds emanating from virtual objects, such as the hum of machinery or footsteps within a virtual factory [124]. In DT interactions, sound helps to reinforce spatial awareness. For example, a maintenance engineer could be navigating through a virtual simulation of a factory floor, and the sound system could provide audio cues about the position of machines or provide alerts related to operational anomalies [174]. These audio cues help users maintain situational awareness without needing to rely solely on visual cues, enhancing the overall interaction with the virtual environment.

- **Multi-touch Smart Screens**: Multi-touch screens, such as the Microsoft Surface Hub, allow for collaborative interaction with DTs. These screens support multi-user touch interfaces, enabling users to simultaneously interact with and manipulate DT models in real-time. This functionality is particularly useful in scenarios where multiple stakeholders are involved in decision-making processes, such as in design, analysis, and troubleshooting [124]. For example, in a smart city DT service, urban planners can gather around a multi-touch screen to interact with a DT of the city, zooming in to explore specific buildings, traffic systems, or public services [60]. They can adjust parameters, simulate changes, and visualize the impacts of different urban planning decisions [174]. This interaction is enriched by the tactile feedback and the ability for multiple users to interact simultaneously, making the process more collaborative and dynamic.
- **Multi-modal Sensor Fusion Chip**: These chips enable real-time sensor data processing by combining inputs from various modalities, including vision, depth sensing, and motion. Devices such as NVIDIA Jetson and Intel RealSense are capable of processing data from multiple sources, such as cameras, microphones, and accelerometers, and synthesizing this information into a coherent understanding of the environment. In DT service interaction, this fusion of data allows for accurate tracking of user movements and environmental conditions [41]. For instance, a worker in a warehouse could wear a smart device with these sensor fusion chips to interact with a DT of the warehouse layout. As the worker moves around, the system can track their location and provide context-sensitive information or guidance, such as real-time inventory levels or potential hazards. By fusing different sensor data, these systems can accurately map user actions and contextualize them within the virtual environment, ensuring smooth and adaptive interaction [88].
- **Human Action Recognition AI Module**: Human action recognition modules, such as OpenPose and MediaPipe, analyze real-time human motion to recognize specific gestures or actions. These AI modules are instrumental in recognizing how users interact with the system, allowing for hands-free control in DT service interactions. For example, in a medical DT model, a surgeon can use hand gestures to manipulate a 3D model of a patient's anatomy without touching any devices [119]. These modules enhance user interaction by enabling natural, intuitive gestures for control and manipulation. In DT service applications, such as remote maintenance of industrial equipment, technicians can use specific hand movements to zoom in on DT models or highlight areas that require attention. This creates a more seam-

less interaction by reducing the reliance on traditional input devices like keyboards or mice [132].

- **Environment and User State Fusion Algorithm**: Transformer-based sensor fusion algorithms utilize advanced machine learning models to process and integrate sensor data from diverse sources. These algorithms are capable of learning complex relationships between user actions and environmental variables, making them ideal for adaptive DT service interactions [67]. In a smart home environment, for example, the system can monitor various sensor inputs, including temperature, lighting, and user movements, to create a personalized experience. A user entering a room may trigger the system to adjust lighting, temperature, and even suggest a service based on their historical preferences. The system's ability to learn from real-time data and adapt to the user's context is a crucial aspect of modern DT service interactions [81].

Enabling Technologies for Universal Device Integration Frameworks: Universal device integration frameworks provide the underlying infrastructure that enables diverse hardware devices to communicate seamlessly within a DT service interaction. These frameworks allow users to interact with a wide range of devices, from Internet of Things (IoT) sensors to complex machinery, in a unified and coherent manner [77].

- **IoT Device Access Standards**: Standards like Matter and Zigbee facilitate the integration of various IoT devices into the DT ecosystem. These protocols enable seamless communication between different smart devices, regardless of manufacturer, creating a unified network for DT service interaction [7]. In a smart city, for instance, Matter could be used to link traffic sensors, streetlights, and surveillance cameras into a common platform, enabling real-time monitoring and control. Users can interact with the DT of the city via a centralized interface, adjusting parameters like traffic light timings or energy consumption based on real-time sensor data [101, 115].
- **Device Abstraction Layer Middleware**: Device abstraction layers like OpenHAB and Home Assistant provide a unified interface for controlling diverse IoT devices in a DT model. These platforms allow users to manage and configure devices without worrying about compatibility or communication protocols [43]. In DT services, this middleware simplifies interaction by abstracting away the complexities of device integration. For example, in a smart building, users can interact with DT of the building's energy system, controlling HVAC, lighting, and security systems through a single interface, regardless of the underlying hardware [121].
- **Cross-Platform Hardware Integration API**: Cross-platform APIs like WebUSB and WebBluetooth enable direct communication with hardware devices over the web, allowing for remote interaction with IoT devices and sensors in a DT service interaction. These APIs enable devices to be controlled from a browser or mobile app, streamlining the user experience and enhancing the interaction process [83]. For instance, a user could use their mobile device to interact with a wearable sensor in a smart healthcare environment, monitoring vital signs in real-time [141]. Web-

Bluetooth could be used to synchronize the data with a cloud-based DT, providing actionable insights for healthcare professionals during consultations [36].

Enabling Technologies for Real-Time Multimedia Event Processing Engines: Real-time multimedia event processing engines are critical for managing high-volume, low-latency data in DT environments. These engines ensure that user actions, sensor data, and other events are processed and delivered to the appropriate systems instantaneously, allowing for dynamic, real-time interactions with DTs [172].

- **Real-time Messaging Bus**: Real-time messaging buses like Apache Kafka, RabbitMQ, and MQTT are fundamental for enabling communication between distributed systems in DT environments. They provide asynchronous messaging and ensure that event streams can be handled efficiently, even in high-throughput scenarios. These systems allow for real-time data delivery to various components of a DT service, including data visualizations, user interfaces, and analytic engines [131]. In DT service interactions, these messaging systems enable the flow of real-time information between sensors, user interfaces, and backend systems. For example, in a smart manufacturing scenario, machine sensors might generate real-time data about machine performance, which is immediately transmitted via Kafka to a DT model for real-time diagnostics and visualization on a user's interface. This seamless and uninterrupted flow of data is essential for enabling fast decision-making and proactive actions in DT models [114].
- **Event Stream Processing Engines**: Event stream processing engines like Apache Flink and Streamlit allow for the real-time analysis of streaming data. These engines process data as it is ingested, enabling continuous insights and decision-making based on the latest data from sensors or other input devices. Apache Flink is particularly well-suited for complex event processing (CEP), allowing it to handle large volumes of data with low latency, while Streamlit focuses on easy-to-deploy dashboards for visualizing event-driven data [50]. In the context of DT service interaction, these tools process streaming data from various sources, whether it's sensors capturing the movement of a physical object, user actions interacting with a DT, or external data sources providing environmental context. Real-time analytics are displayed instantly to users, ensuring that interactions are based on the most current data available, which is particularly important in industries such as healthcare, manufacturing, and energy management [113].
- **Low-latency Multimodal Interaction Framework**: The Unity XR Interaction Toolkit is a low-latency multimodal interaction framework that allows developers to create immersive experiences by integrating physical input with virtual elements. Extended Reality (XR) encompasses both AR and VR, and this toolkit provides a framework for users to interact with DTs via touch, gestures, and even voice commands in real-time [64]. In DT services, Unity XR enables users to interact with DTs of physical assets in a highly responsive manner. For example, a user in a VR simulation of a manufacturing plant can perform real-time maintenance on virtual machines by using hand gestures to manipulate equipment or touch to navigate through control panels. Unity XR ensures that these interactions

are processed with minimal delay, creating a smooth and engaging experience that blends both virtual and physical elements [8].

Enabling Technologies for Intelligent Context-Aware Service Orchestration: Intelligent context-aware service orchestration enables the dynamic adjustment of service workflows based on real-time data, user context, and environmental conditions. This technology ensures that DT services can adapt to the evolving needs of users, providing context that is specific information and actions [140].

- **Context-aware Platform**: Context-aware platforms, powered by technologies like Contextual AI and the Google Awareness API, allow DT models to understand the user's context and environment in real time. These platforms use sensors, GPS, and other data inputs to determine the user's location, activities, and preferences, adapting the DT model's response accordingly [74]. For instance, in a smart retail environment, the context-aware platform could recognize that a customer is near a specific product in a store and, based on past behavior, offer personalized promotions via a mobile app. This seamless interaction, powered by contextual awareness, enables dynamic and responsive DT services that cater to user needs as they change in real-time [167].
- **Intelligent Service Orchestration Engine**: Service orchestration engines like Camunda BPMN and AWS Step Functions manage the execution of complex workflows by coordinating the various services and data components involved in a DT model. These engines ensure that services are executed in the correct sequence, based on real-time data and user input [99]. In a DT service interaction, these orchestration engines allow for complex scenarios, such as remote diagnostics in industrial settings, where multiple backend services must be coordinated. If a user encounters an issue with a piece of machinery, the orchestration engine can trigger a sequence of actions, including accessing the machine's DT, checking maintenance logs, and triggering alerts for remote engineers [45].
- **Scenario Switching and Service Routing**: Scenario switching and service routing technologies like IFTTT and Node-RED allow users to define triggers and automate actions in response to specific conditions. These tools provide users with the flexibility to create customized workflows based on changing environmental or contextual factors [78]. For example, in a smart home DT service, a user might set up an IFTTT scenario where the thermostat adjusts based on the weather forecast, and the lights dim when they are watching a movie. Node-RED can be used to create more complex automation sequences, integrating multiple sensors and services to optimize the smart home experience [98].

Enabling Technologies for Real-Time Collaborative Interaction Platforms: Real-time collaborative interaction platforms enable multiple users to interact simultaneously with DTs, providing a shared, synchronized experience across locations and devices. These platforms support collaborative decision-making, design, and monitoring in DT environments [168].

- **Real-time Collaborative Editing Systems**: Real-time collaborative editing systems, such as Firebase Realtime Database and frameworks that are based of Conflict-free Replicated Data Type (CRDT), allow users to interact with and modify shared DTs simultaneously. These systems ensure that all users' changes are reflected in real time, enabling collaborative tasks such as co-designing, troubleshooting, and analyzing data in a synchronized manner [59]. In a DT service for urban planning, multiple architects could interact with the DT of a city's infrastructure simultaneously, making changes to different parts of the model while receiving instant feedback on the impact of their modifications. The system ensures that all users have an up-to-date version of the model, preventing conflicts and allowing for seamless collaboration [116].
- **Distributed State Synchronization**: Distributed state synchronization technologies like WebRTC and Operational Transformation ensure that multiple users' actions are synchronized across devices, maintaining consistency in the shared DT. These technologies allow for real-time communication and interaction, even across different networks and platforms [104]. For example, in a virtual collaboration session for a smart factory DT, two engineers could remotely inspect different parts of the factory floor, making real-time annotations and adjustments to the DT. WebRTC ensures that their actions are immediately reflected for all participants, while Operational Transformation ensures that changes are consistently synchronized, regardless of network conditions [109].
- **Multi-user Permissions and Session Management**: Multi-user permissions and session management tools like OAuth2.0 and Socket.io Rooms manage user access and ensure that each participant in the DT service interaction has the appropriate level of access to data and features. These tools support collaborative workflows while ensuring data security and user privacy [68]. In a collaborative DT service for a research project, different team members might have different roles, such as a project manager, engineer, and data analyst. OAuth2.0 allows for secure, role-based authentication, ensuring that only authorized users can make specific changes to the DT. Socket.io Rooms enables real-time communication and updates among users within a session, keeping all participants informed and synchronized during their interactions [96].

(4) *DT Evolution-Aware Wireless Communication Technologies*

The preceding analysis has revealed a stringent set of performance objectives for requirements and challenges in wireless networks that must simultaneously ingest high volume sensory data to construct and update DTs and then sustain immersive, low latency interactions between the physical and virtual worlds. These objectives cannot be met by incremental improvements of legacy radio stacks; instead, they demand the systematic integration of emerging physical layer, MAC, network and cross-domain technologies that jointly shape the propagation environment, the spectral efficiency and the end-to-end service guarantees. In this subsection, we review the most salient enabling technologies, describing for each one the concrete mech-

Core Technology	Function	Solved Problem(s)
RIS	Dynamically shapes wireless propagation paths via tunable metamaterial elements.	**DT Construction:** Stabilizes uplink SINR for lossless data ingestion; enables elastic capacity during transient events. **DT Service Interaction:** Ensures inter-modal synchronization for high-fidelity downlink.
Semantic Communication	Transmits task-relevant semantic features instead of raw data using AI encoders/decoders.	**DT Construction:** Reduces uplink traffic; enables energy-proportional transmission. **DT Service Interaction:** Compresses interaction requests for uplink; reconstructs multimodal feedback.
Terahertz Communications	Leverages ultra-broadband spectrum for extreme data rates.	**DT Construction:** Delivers multi-Gbps downlink; multimodal feedback streams **DT Service Interaction:** enables dynamic QoS slicing; minimizes inter-modal skew via spatial isolation.
Cell-Free Massive MIMO	Distributes coherent processing across APs; central unit handles precoding/decoding.	**DT Construction:** Provides macro-diversity for stable uplink; reduces sensor transmit power. **DT Service Interaction:** Broadcasts precision timing beacons for synchronization.
Zero-Trust Security + PUFs	Uses hardware-unclonable keys for authentication; enforces ephemeral credentials.	**DT Construction:** Prevents compromised sensors from poisoning model calibration. **DT Service Interaction:** Enables seamless zero-trust mobility; resists inference attacks.
LEO Satellite Offloading	Provides global coverage via satellites with regenerative payloads/edge processors.	**DT Construction:** Ensures lossless data ingestion in remote/mobile scenarios; enables energy-aware scheduling. **DT Service Interaction:** Extends deterministic interaction to rural areas.
TSN	Integrates Layer 2 deterministic guarantees into 5G/6G UPF.	**DT Construction:** Ensures elastic capacity via frame preemption. **DT Service Interaction:** Enforces deterministic uplink for interaction requests.
NOMA	Superposes users in power domain; decodes via SIC.	**DT Construction:** Scales medium access for dense sensor arrays. **DT Service Interaction:** Dynamically prioritizes latency-sensitive interaction traffic via power allocation.

Fig. 2.7 DT evolution-aware wireless communication technologies

anisms through which it mitigates the identified requirements and challenges. An overview of all corresponding technologies is illustrated in Fig. 2.7.

- **Reconfigurable Intelligent Surfaces (RIS) for Programmable Propagation**: A reconfigurable intelligent surface is a planar array of sub-wavelength metamaterial elements whose electromagnetic response can be tuned in real time through low power control signals. By adjusting the phase, amplitude, and in some cases the polarisation state of each element, an RIS can synthesise anomalous reflec-

tion, refraction, and focusing patterns that effectively create electronically steerable beams between transmitters and receivers without additional radio frequency chains. The surface acts as a passive or semi passive relay; it does not regenerate the signal but shapes the propagation channel itself, thereby offering spatial gain, interference suppression, and coverage extension with minimal added noise or power consumption.
For the uplink data ingestion task, RIS deployment around critical assets produces two primary benefits. First, by concentrating energy along deterministic paths, the array raises the effective signal to interference plus noise ratio (SINR) for sensor nodes located in deep shadow or behind metallic clutter [49]. This gain allows the radio interface to maintain a sufficiently low packet loss rate without resorting to aggressive modulation or high transmit power, thus contributing to the energy proportional transmission mandate. Second, the spatial focusing property mitigates large scale fading excursions, reducing the instantaneous dynamic range of the channel and enabling the scheduler to predict achievable data rates with higher confidence [166]. This predictability is essential for the elastic capacity requirement that the network expand or contract within milliseconds in response to transient data surges. In the downlink direction, an RIS can establish multiple independent beams that serve concurrent users without mutual interference [16]. Each beam can be shaped to deliver the exact power distribution required to meet the high fidelity multimedia throughput target while simultaneously ensuring that intermodal skew remains within the bound. The phase coherent nature of the surface allows per subcarrier delay pre compensation, which flattens the group delay response across the entire 5 GHz instantaneous bandwidth envisioned. Synchronisation is further enhanced because the RIS itself introduces negligible processing delay compared with active relays.
In addition, RIS can precisely spatially resolve interference suppression in dense deployments. When thousands of sensors are distributed over a 1 km^2 construction domain, co channel interference becomes the dominant impairment. An RIS can create spatial nulls toward unintended receivers while maintaining constructive gain toward the intended edge server without requiring orthogonal spectrum partitions [165].

- **Semantic Communication for Task-Oriented Bit Transport**: Semantic communication represents a paradigm shift in wireless data transmission, moving beyond the traditional Shannon-centric goal of perfect bit-level fidelity [84]. Instead, it focuses on conveying only the task-relevant information required by the receiving DT or user application. This is achieved through learned encoders at the transmitter that extract high-level semantic features from raw source data. These features are transmitted as compact representations over the radio channel, enabling the receiver to reconstruct the intended meaning within the fidelity demanded by downstream decision functions. This approach is exceptionally well-suited to DT model, where raw sensor data is ultimately reduced to a small set of physical parameters, state estimates, or control actions for model calibration or user interaction.
For the continuous and lossless data ingestion requirement critical to DT construction, semantic communication drastically alleviates network load. Semantic

codecs reduce instantaneous uplink traffic by one to two orders of magnitude while preserving the information essential for accurate model calibration and update. For instance, a dense LiDAR point cloud comprising millions of spatial coordinates can be distilled by a semantic encoder into a few hundred compact descriptors capturing high-level features like surface curvature, edge orientation, and occupancy probability [54]. This payload reduction relaxes the peak rate requirement from the challenging huge envelope to a more manageable few hundred Mbps. Consequently, the MAC layer can sustain the stringent packet loss target with less aggressive channel coding and lower retransmission overhead. Additionally, shorter semantic packets experience reduced queuing delays and faster decoding times, directly improving the latency budget and supporting near-deterministic delivery.

Semantic communication also intrinsically enables energy-proportional transmission, addressing a core challenge in resource-constrained sensing nodes [163]. The length of the semantic codeword emitted by the encoder naturally adapts to the information entropy of the physical scene. During quiescent operational periods, semantic entropy is low, resulting in very short codewords that minimize radio airtime and energy consumption. Conversely, during critical transient events, semantic entropy rises, triggering longer codewords that ensure high-value information is fully conveyed without loss. Crucially, the energy expended per transmission now scales directly with the semantic value of the data rather than the raw sample count or bit rate. This dynamic alignment satisfies the energy proportionality mandate without introducing complex protocol-level negotiations or explicit value signaling from DT, simplifying model design.

In the context of DT service interaction, semantic communication enhances both uplink determinism and downlink efficiency. For uplink interaction requests semantic encoders compress user intent (voice, gestures, commands) into ultra-compact embeddings, shrinking payloads to mere tens of bytes. This reduction minimizes packet airtime, making the stringent 10 ms end-to-end deterministic latency bound achievable even under variable channel conditions. On the downlink, semantic decoders reconstruct high-fidelity multimodal feedback from lightweight latent codes transmitted over the air. This allows the network to meet the peak 5 Gbps per-user data rate requirement while consuming significantly less spectrum compared to transmitting raw or conventionally compressed data. Freed spectral resources can then support more concurrent users or enhance robustness through redundancy. The shared semantic knowledge base periodically updated between endpoints, ensures that both intent and feedback are reconstructed with the context-aware fidelity demanded by immersive interaction.

- **Terahertz Communications for Ultra-Broadband Downlink Delivery**: Terahertz (THz) communication systems emerge as a pivotal enabling technology specifically addressing the stringent high-fidelity downlink requirements essential for immersive DT service interaction. The fundamental advantage of THz lies in its access to vast contiguous bandwidths, often on the order of tens of gigahertz [1]. This unprecedented spectral resource directly translates to the potential for peak data rates exceeding 100 Gbps. Such capacity is indispensable for the simultaneous, lossless delivery of the multimodal feedback streams demanded by interactive DT applications. Crucially, a single contiguous THz channel allocation, can easily support throughput exceeding the peak of the physical layer [24, 25, 170]. This comfortably surpasses the aggregate 5 Gbps per user target mandated for high-fidelity downlink, even after accounting for necessary forward error correction overhead, thereby providing significant headroom for reliable operation under peak load conditions.

Furthermore, the physics of THz propagation significantly aids in overcoming the critical challenge of maintaining inter-modal synchronisation errors below 200 microseconds for concurrent visual, auditory, and haptic flows. The extremely short wavelengths at THz frequencies enable the practical realization of massive antenna arrays integrated into compact form factors. These arrays generate highly directional beams with inherently narrow beamwidths. This spatial precision allows access points to form numerous simultaneous, spatially separated, and effectively non-interfering beams directed towards multiple users within dense deployment scenarios like factories or transportation hubs. By isolating user streams in the spatial domain, THz communications drastically reduce co-channel interference and the associated jitter and path delay variations that plague synchronisation efforts at lower frequencies. This inherent spatial separation capability directly satisfies the spatial multiplexing demands of dense interactive deployments outlined in the scalability challenge for DT service interaction.

The ultra-wide bandwidth intrinsic to THz links also provides a powerful mechanism to address the requirement for dynamic QoS negotiation under heterogeneous user demands. The abundance of spectrum enables highly flexible and fine-grained resource slicing in the frequency domain. Each distinct slice of the available THz bandwidth can be dynamically assigned to a specific user session and configured with an independent modulation and coding scheme (MCS) [12], tailored precisely to that session's instantaneous latency, reliability, and throughput requirements. For example, a remote operator engaged in precision teleoperation might be allocated a narrower slice configured for ultra-low latency using robust, lower-order modulation, prioritizing command responsiveness over raw throughput. Conversely, a quality inspector performing a high-resolution model review could receive a wider slice utilizing high-order modulation to maximize fidelity and detail, accepting slightly higher latency if necessary. Critically, this resource allocation and MCS adaptation can occur dynamically, even within the duration of a single symbol period, based on real-time feedback regarding user intent and application context. This fine-grained control significantly simplifies the complex optimization problem posed by the need to simultaneously support diverse and evolving latency-fidelity trade-offs across numerous concurrent DT interaction sessions.

- **Cell Free Massive MIMO with Distributed Coherent Processing**: Cell free massive MIMO replaces the conventional high power macro base station with a large number of geographically distributed access points that jointly serve all users via coherent signal processing. All access points are connected through a high capacity fronthaul network to a central processing unit that performs joint channel estimation, precoding and decoding. The resulting system behaves like a single colossal antenna array whose aperture spans hundreds of metres, offering uniform coverage, macro diversity, and high spatial multiplexing gain without cell edge boundaries.

For DT construction, the stable uplink channel quality enabled by macro diversity ensures continuous, lossless ingestion of sensor data into DT [11]. Independent small scale fading at each distributed antenna is averaged out at the central processor, yielding a nearly deterministic effective channel. This stability allows the link adaptation algorithm to operate with a much smaller backoff margin, which in turn increases the instantaneous spectral efficiency and supports the 1 Mbps to 5 Gbps elastic range without manual reconfiguration. This is essential for high-fidelity DT model updates. The distributed architecture also reduces the probability of simultaneous deep fades across all antennas, thereby maintaining the low user plane packet loss ratio even under severe shadowing.

For DT service interaction, distributed antennas act as precise timing beacons whose signal is received at microsecond level accuracy by nearby sensors. By leveraging the known geometry of the distributed array, sensors can perform trilateration based time transfer, achieving the required offset bound without the need for high power GPS receivers or wired synchronisation infrastructure. The central processor can also broadcast a common time reference over the coherent downlink, further tightening the synchronisation bound.

Energy proportional transmission is facilitated because the spatial focusing gain of cell free massive MIMO allows sensors to reduce transmit power while still meeting the SINR target [123]. The precoding vectors computed at the central unit can be designed to minimise the total uplink transmit energy subject to individual reliability constraints, effectively embedding the energy value metric into the DT interaction sessions.

- **Zero-Trust Security Framework with Physically Unclonable Functions**: Embedded Physically Unclonable Functions (PUFs) establish an unforgeable hardware root-of-trust for DT model, addressing unmitigated challenges in device authentication and data provenance. PUFs exploit nanoscale manufacturing variations in sensor silicon to generate unique cryptographic fingerprints that are physically impossible to clone or predict [29]. When integrated with lightweight zero-trust architectures, they enable continuous attestation of every sensing node and interaction device without centralized certificate authorities. This directly satisfies the unlinkability requirement while eliminating authentication latency, which is a critical gap in mobility scenarios where traditional PKI handshakes exceed 10 ms deadlines.

  For DT construction, after collecting the feature data, physical layer security (PLS) can be employed to protect the wireless signals [69, 169], while the PUF-based authentication, can prevent compromised sensors from poisoning model calibration. Each vibration sensor embedded in a turbine blade cryptographically signs its measurements using keys derived from its unique PUF response. Edge verifiers validate these signatures within microseconds, ensuring only authenticated data contributes to DT,which is crucial for maintaining the lossless ingestion integrity requirement against false data injection attacks. Furthermore, PUF keys are dynamically regenerated from environmental noise, creating ephemeral credentials that satisfy privacy preservation without computational overhead.

During DT service interaction, PUFs enable seamless zero-trust mobility. As an engineer moves from a private 5G network to a macro cell, their haptic controller continuously reauthenticates using PUF-derived session keys negotiated via lattice-based cryptography. This occurs without backend involvement, reducing handover latency to <2 ms while maintaining the strict determinism required for teleoperation. This framework also introduces adaptive trust scoring. The behavioral machine learning model for analyzing interaction patterns on edge nodes can dynamically adjust the authentication frequency, reduce encryption overhead during regular operations, and enhance the review intensity when abnormal operations are detected.

The technology fundamentally transforms privacy preservation through hardware-enforced data compartmentalization. Sensitive operational data from different DT sub-models are encrypted with distinct PUF-derived keys accessible only to authorized DT components. Crucially, PUF responses generate unique RF fingerprints for each transmitter, allowing physical-layer identification that thwarts impersonation attempts during spectrum degradation events.

- **Joint Communication and Computation Offloading over LEO Satellite Constellations**: Low Earth orbit satellite networks equipped with regenerative payloads and on board edge processors create a global wireless fabric that complements terrestrial links. Each satellite carries a software defined radio capable of adaptive beamforming, a GPU cluster for virtualised baseband processing, and an inter satellite optical backbone. The architecture enables direct radio links between sensors or user devices and satellites at elevations above 500 km, providing line of sight connectivity even in rural, maritime, or airborne scenarios.

  For DT construction, LEO satellites eliminate terrestrial coverage gaps for continuous, lossless data ingestion from remote assets. The regenerative payload demodulates uplink packets on board, applies advanced LDPC and polar codes, and retransmits only the necessary parity segments when terrestrial gateways are unavailable. This store carry forward mechanism guarantees that the packet loss ratio remains low level even during multi second gateway outages. Furthermore, the constellation's orbital motion provides macro diversity: when one satellite disappears beyond the horizon, a trailing satellite already has the user in its footprint, ensuring uninterrupted ingress without handover latency.

  For DT service interaction, this technology can also assist in adjusting the energy ratio. Satellites broadcast downlink ephemeris and channel state information that allow ground sensors to predict link availability and adjust their duty cycle accordingly [148]. A sensor can postpone transmission until the satellite elevation angle exceeds a threshold that minimises path loss, thereby reducing transmit energy while still meeting the latency budget. The on board scheduler can also aggregate packets from multiple sensors into single bursts, amortising the fixed energy cost of radio wake up across many samples.

- **Time Sensitive Networking (TSN) over 5G/6G Integrated Access and Backhaul**: TSN provides a critical foundation for deterministic communication essential to both DT construction and service interaction by offering Layer 2 guarantees of bounded latency, ultra-low jitter, and reliable packet delivery. Its integration into the 5G/6G User Plane Function (UPF) creates a seamless, unified deterministic transport service that bridges the traditionally separate domains of radio access network (RAN) and backhaul/fronthaul networks [52]. This holistic approach ensures that latency-critical traffic generated by DT processes—whether uplink sensor data streams for model construction/update or bidirectional interaction commands and feedback—receives predictable and prioritized treatment consistently across the entire path, from the sensor/device antenna to the computational node hosting the DT and back.
  The stringent requirement for the network to expand its capacity within milliseconds during transient events in DT construction, as outlined in the challenges, is directly met by TSN's frame preemption capability. When a burst of high-priority DT construction data arrives at any TSN-capable switch or router along the path, it can immediately interrupt the transmission of lower-priority background traffic. This eliminates queue build-up for critical data virtually instantaneously. Furthermore, when combined with TSN's Credit-Based Shaping (CBS), this mechanism guarantees that the queuing delay contribution per network hop remains tightly bounded [157]. This predictable per-hop behavior is crucial because it ensures the overall end-to-end latency envelope remains manageable and within required bounds even during the most extreme and sudden load spikes associated with DT data ingestion surges, fulfilling the "elastic capacity" mandate without manual reconfiguration.
  The critical need for deterministic uplink delivery of interaction requests, with bounded latency below 10 ms and negligible jitter and packet loss as defined in the service interaction challenges, is fundamentally enabled by the deep integration of TSN within the 5G/6G data plane. This integration involves mapping specific 5G Radio Bearers carrying high-priority interaction traffic to dedicated, high-priority TSN traffic classes at the UPF. The gNodeB (base station) schedules uplink radio resources for these bearers with awareness of the subsequent TSN path. Crucially, once the data leaves the RAN and enters the backhaul as Ethernet frames tagged with their TSN class, every TSN bridge in the path enforces strict priority scheduling based on this class. This guarantees that the variable delays often introduced by conventional backhaul networks are eliminated, preventing them from consuming the strictly limited end-to-end latency budget allocated for user interaction responsiveness.
- **Non Orthogonal Multiple Access (NOMA) with Successive Interference Cancellation**: NOMA fundamentally rethinks resource allocation in wireless networks by superposing multiple user signals within the same time-frequency resource block. Unlike traditional orthogonal schemes, NOMA leverages power-domain multiplexing to assign distinct power levels to users based on their instantaneous channel conditions and QoS requirements [171]. At the receiver, Successive Interference Cancellation (SIC) is employed to decode the strongest signal first, subtract

its contribution from the aggregate received signal, and iteratively decode weaker signals. This approach eliminates the need for strict orthogonality, dramatically improving spectral efficiency and scalability while directly addressing several critical challenges in DT deployments.

For scalable medium access in massive sensor arrays, NOMA proves exceptionally effective [30]. By permitting A large number of sensors to share a single physical resource block, NOMA overcomes the limitations of both grant-based (signaling overhead) and grant-free (multi-user interference) schemes. The SIC process dynamically adapts to channel conditions: the decoding order is determined by instantaneous received power levels, ensuring sensors with stronger channels are decoded first without explicit coordination. This emergent intelligence allows the system to scale to densities exceeding 10,000 sensors per km^2 under bursty traffic, as heterogeneous data streams from visual, thermal, or vibration sensors coexist without dedicated spectral partitions. The scheme's inherent adaptability ensures reliability targets are met even during transient data surges, fulfilling the "elastic capacity" requirement for DT construction.

NOMA further revolutionizes dynamic QoS negotiation under heterogeneous user demands. Its power allocation framework can be reconfigured on a per-slot basis to prioritize latency-sensitive or fidelity-critical interactions. For instance, during DT service interaction, a teleoperation command may be assigned high power to ensure immediate decoding at the SIC receiver, while a high-resolution model inspection stream operates at lower power but benefits from interference cancellation in later SIC stages [90]. This optimization, executed centrally at the baseband processor and signaled via downlink control channels, achieves sub-millisecond resource reallocation. Consequently, the network dynamically aligns with evolving user intent–such as shifting from precision manipulation to quality review–without manual intervention.

2.3 Summary of the Chapter

This chapter comprehensively outlines the key issues and enabling technologies for deploying DT over wireless networks, structuring the discussion around three core pillars: DT construction, DT service interaction, and the wireless network infrastructure. The initial section details the significant design requirements and challenges, establishing that DT construction demands high-fidelity data from real-time multimodal sensors, supported by robust hardware and software capable of synchronous collection, fusion, and scalable storage. Key challenges include ensuring data integrity, temporal consistency across heterogeneous sources, and managing massive datasets efficiently. For DT service interaction, the focus shifts to delivering immersive, low-latency, and collaborative user experiences through hardware like AR/VR headsets and software that can manage real-time multimodal inputs and adapt interfaces dynamically. A fundamental challenge is synchronizing multi-sensory feedback and supporting seamless multi-user collaboration without compromising data consistency. The chapter argues that conventional wireless networks are ill-equipped for these tasks, necessitating a paradigm shift from best-effort statistical performance to deterministic service guarantees. This involves supporting both the high-volume, reliable uplink for DT construction (e.g., continuous data ingestion, synchronized sampling) and the ultra-low-latency, high-fidelity bidirectional flows for DT service interaction (e.g., deterministic interaction requests, multimodal feedback).

To address these multifaceted challenges, the chapter presents a suite of enabling technologies. For DT construction, hardware advancements such as LiDAR, IMUs, and GNSS provide precise data acquisition, while Edge AI accelerators and gateways enable localized processing to reduce latency and bandwidth consumption. On the software side, time synchronization protocols like PTP/NTP, middleware messaging systems including DDS and Kafka, and distributed storage like HDFS are crucial for managing data flow and consistency. Modeling and dynamic updates are powered by simulation engines like Unity3D and Unreal Engine, with security being a foundational layer addressed by encryption and role-based access control. For DT service interaction, hardware such as AR/VR headsets and holographic displays creates immersive interfaces, complemented by AI-powered sensor fusion and distributed spatial computing to ensure responsive, context-aware engagement. Software technologies for interaction include real-time event processing engines, advanced multimodal fusion algorithms to interpret user intent, and collaborative platforms like Firebase to enable synchronized multi-user experiences. Critically, the chapter introduces a host of advanced wireless technologies to build the necessary deterministic network fabric. These include RIS to shape the wireless channel, Semantic Communication to transmit only task-relevant information, and Cell-Free Massive MIMO to provide uniform, reliable coverage. Other key technologies are ISAC for predictive channel state information, HFL for efficient, privacy-preserving model training at the edge, and Zero-Trust Security with PUFs to establish an unforgeable hardware root-of-trust.

Several key lessons can be learned from this comprehensive analysis. First, the transition to functional DTs necessitates a fundamental shift in network design philosophy from simply moving bits to enabling task-oriented, deterministic communication, where performance is measured by the success of the application (e.g., model accuracy, interaction fidelity). Second, a holistic, cross-layer design is not merely beneficial but essential. Hardware, software, and networking technologies are deeply interdependent and must be co-designed to meet the stringent end-to-end requirements. For instance, edge computing hardware is only effective when supported by low-latency wireless links and intelligent software for data processing and model updates . Third, intelligence must be distributed and moved to the network edge. Technologies like HFL and edge AI are paramount for managing the immense data volumes and latency constraints, enabling real-time analytics and personalization without overwhelming centralized resources. Finally, security and privacy cannot be afterthoughts. The adoption of frameworks like Zero-Trust Security, enforced by hardware anchors like PUFs, illustrates that trust must be verifiably established at every point in the system, from sensor to cloud, to ensure the integrity of the DT and the safety of its interactions.

References

1. Akyildiz, I., Han, C., Hu, Z., Nie, S., Jornet, J.: Terahertz band communication: an old problem revisited and research directions for the next decade. IEEE Trans. Commun. **70**, 4250–4285 (2021)
2. Akyildiz, I., Lee, W.Y., Vuran, M., Mohanty, S.: Next generation/dynamic spectrum access/cognitive radio wireless networks: a survey. Comput. Netw. **50**, 2127–2159 (2006)
3. Alja'afreh, M., Al Mallah, R., Karime, A., El Saddik, A.: Cybersecurity in the metaverse: challenges and approaches. In: 2023 International Conference on Intelligent Metaverse Technologies & Applications, pp. 1–8 (2023)
4. Alkhateeb, A., Jiang, S., Charan, G.: Real-time digital twins: vision and research directions for 6G and beyond. IEEE Commun. Mag. **61**(11), 128–134 (2023)
5. Anchitaalagammai, J., Kavitha, S., Buurvidha, R., Santhiya, T.S., Roopa, M.D., Sankari, S.S.: Edge artificial intelligence for real-time decision making using NVIDIA Jetson Orin, Google coral edge TPU and 6G for privacy and scalability. In: 2025 International Conference on Visual Analytics and Data Visualization, pp. 150–155 (2025)
6. Apthorpe, N.J., Reisman, D., Sundaresan, S., Narayanan, A., Feamster, N.: Spying on the smart home: privacy attacks and defenses on encrypted IoT traffic. Comput. Res. Repository (2017). arXiv:1708.05044
7. Belli, D., Barsocchi, P., Palumbo, F.: Connectivity standards alliance matter: state of the art and opportunities. IOT **25**, 101005 (2024)
8. Bhaskaran, V., Mahbub, U.: Immersive user experiences: trends and challenges of using XR technologies. Comput. Vis., 260–278 (2024)
9. Bi, S., Yuan, X., Hu, S., Li, K., Ni, W., Hossain, E., Wang, X.: Failure analysis in next-generation critical cellular communication infrastructures. Comput. Res. Repository (2024). arXiv:2402.04448
10. Bilal, M., Ali, G., Iqbal, M.W., Anwar, M., Malik, M.S.A., Kadir, R.A.: Auto-prep: efficient and automated data preprocessing pipeline. IEEE Access **10**(000), 21 (2022)
11. Boudreau, G., Srinivas, K.V., Shahbazpanahi, S., Adve, R.S., Ammar, H.A.: User-centric cell-free massive MIMO networks: a survey of opportunities, challenges and solutions. IEEE Commun. Surv. & Tut. **24**, 611–652 (2021)
12. Boulogeorgos, A.A.A., Papasotiriou, E.N., Alexiou, A.: A distance and bandwidth dependent adaptive modulation scheme for THZ communications. In: 2018 IEEE International Workshop on Signal Processing Advances in Wireless Communication, pp. 1–5 (2018)
13. Bărbuceanu, T., Craciunescu, R., Popovici, E.: Data protection device. In: 2023 15th International Conference on Electronics, Computers and Artificial Intelligence, pp. 01–06 (2023)
14. Cai, Z., Huang, J., Huang, X., Li, Y., Xiong, W., Tang, Z., Liang, B.: An efficient power management circuit for battery monitoring systems. In: 2024 4th International Conference on Electronic Information Engineering and Computer Communication, pp. 204–208 (2024)
15. Cakir, L.V., Al-Shareeda, S., Oktug, S.F., Özdem, M., Broadbent, M., Canberk, B.: How to synchronize digital twins? A communication performance analysis. In: Proceedings of IEEE CAMAD, pp. 123–127. IEEE (2023)
16. Chen, J., Li, Y., Yi, C., Gong, S.: Generative AI-aided QOE maximization for RIS-assisted digital twin interaction. In: Proceedings of INFOCOM WKSHPS. London, United Kingdom (2025)
17. Chen, J., Shi, Y., Yi, C., Du, H., Kang, J., Niyato, D.: Generative AI-driven human digital twin in IoT-healthcare: a comprehensive survey. IEEE Internet Things J. **11**(21), 34749–34773 (2024)
18. Chen, J., Yi, C., Du, H., Niyato, D., Kang, J., Cai, J., Shen, X.: A revolution of personalized healthcare: enabling human digital twin with mobile AIGC. IEEE Netw. **38**(6), 234–242 (2024)
19. Chen, J., Yi, C., Okegbile, S.D., Cai, J., Shen, X.S.: Networking architecture and key supporting technologies for human digital twin in personalized healthcare: a comprehensive survey. IEEE Commun. Surv. Tutor. **26**(1), 706–746 (2024)

20. Chen, J., Yi, C., Wang, R., Zhu, K., Cai, J.: A joint optimization of sensor activation and mobile charging scheduling in industrial wireless rechargeable sensor networks. In: Proceedings of IEEE ICC, pp. 3568–3573 (2022)
21. Chen, J., Yi, C., Wang, R., Zhu, K., Cai, J.: Learning aided joint sensor activation and mobile charging vehicle scheduling for energy-efficient WRSN-based industrial IoT. IEEE Trans. Veh. Technol. **72**(4), 5064–5078 (2023)
22. Chen, R., Yi, C.: A game-theoretic online optimization for federated digital twin construction via wireless sensing. In: Proceedings of IEEE ICC. Montreal, Canada (2025)
23. Chen, R., Yi, C., Zhou, F., Kang, J., Wu, Y., Niyato, D.: Federated digital twin construction via distributed sensing: a game-theoretic online optimization with overlapping coalitions. IEEE Trans. Mob. Comput. pp. 1–17 (2025)
24. Chen, R., Yi, C., Zhu, K., Cai, J., Chen, B.: A DRL-based hierarchical game for physical layer security with dynamic trilateral coalitions. In: Proceedings of IEEE ICC, pp. 4495–4500 (2023)
25. Chen, R., Yi, C., Zhu, K., Chen, B., Cai, J., Guizani, M.: A three-party hierarchical game for physical layer security aware wireless communications with dynamic trilateral coalitions. IEEE Trans. Wirel. Commun. **23**(5), 4815–4829 (2024)
26. Chen, X., Zhu, W., Chen, J., Zhang, T., Yi, C., Cai, J.: Edge computing enabled real-time video analysis via adaptive spatial-temporal semantic filtering. In: IEEE International Conferences on Internet of Things (iThings) and IEEE Green Computing & Communications (GreenCom) and IEEE Cyber, Physical & Social Computing (CPSCom) and IEEE Smart Data (SmartData) and IEEE Congress on Cybermatics (Cybermatics), pp. 262–267 (2023)
27. Chen, Y.M., Chen, C.J., Chao, Y.L.: A master-slave distributed power management architecture for dynamic voltage and frequency scaling (DVFS) for low power microprocessor. In: 2024 IEEE Applied Power Electronics Conference and Exposition, pp. 1900–1904 (2024)
28. Choi, W.S., Lee, S.J., Kim, J.O., Choi, S.G.: Multicore packet distribution method using multicore network interface card based on tile-gx72 network processor. In: 2024 26th International Conference on Advanced Communications, pp. 350–353 (2024)
29. Delvaux, J., Peeters, R., Gu, D., Verbauwhede, I.: A survey on lightweight entity authentication with strong PUFS. ACM Comput. Surv. **48**(2), 1–42 (2015)
30. Ding, J., Cai, J., Yi, C.: An improved coalition game approach for MIMO-noma clustering integrating beamforming and power allocation. IEEE Trans. Veh. Technol. **68**(2), 1672–1687 (2019)
31. Ding, M., Li, J., Yi, C., Cai, J.: Federated learning for covid-19 on heterogeneous CXR images with noise. In: Proceedings of IEEE ICC, pp. 3413–3418 (2023)
32. Dmitriev, S.O., Valter, D.A., Kontsov, A.M.: System for efficient storage and version control of arbitrary file collections. In: 2020 IEEE Conference of Russian Young Researchers in Electrical and Electronic Engineering (2020)
33. Fingler, H., Ra, M.R., Panta, R.: Scalable, efficient, and policy-aware deduplication for primary distributed storage systems. In: 2019 31st International Symposium on Computer Architecture and High Performance Computing, pp. 180–187 (2019)
34. Gholami, A., Kwon, K., Wu, B., Tai, Z., Yue, X., Jin, P., Zhao, S., Keutzer, K.: Squeezenext: hardware-aware neural network design. In: Proceedings of IEEE CVPRW, pp. 1719–171909 (2018)
35. Giron, A.A.: Data encryption for the TLS protocol against quantum adversaries. Springer, Cham (2025)
36. Grant, R., Rockot, K.: Webusb: Secure and direct usb device access from the web. WICG (2021)
37. Gui, Y.F., Jin, L.F., Ding, D.Z., Xie, Q.L., Wu, S.W., Zou, Y.Q.: Design of X-Ku band broadband high performance and high integration T/R module. In: 2021 13th Global Symposium on Millimeter-Waves & Terahertz, pp. 1–3 (2021)
38. Hao, Y., Qin, X., Chen, Y., Li, Y., Sun, X., Tao, Y., Zhang, X., Du, X.: TS-benchmark: a benchmark for time series databases. In: 37th IEEE International Conference on Data Engineering, pp. 588–599 (2021)

80. Okegbile, S.D., Cai, J., Chen, J., Yi, C.: A reputation-enhanced shard-based byzantine fault-tolerant scheme for secure data sharing in zero trust human digital twin systems. IEEE Internet Things J. **11**(12), 22726–22741 (2024)
81. Okegbile, S.D., Cai, J., Wu, J., Chen, J., Yi, C.: A prediction-enhanced physical-to-virtual twin connectivity framework for human digital twin. In: IEEE Transactions on Cognitive Communications and Networking (2025)
82. Okegbile, S.D., Cai, J., Yi, C., Niyato, D.: Human digital twin for personalized healthcare: vision, architecture and future directions. IEEE Netw. **37**(2), 262–269 (2023)
83. Okegbile, S.D., Cai, J., Zheng, H., Chen, J., Yi, C.: Differentially private federated multi-task learning framework for enhancing human-to-virtual connectivity in human digital twin. IEEE J. Sel. Areas Commun. **41**(11), 3533–3547 (2023)
84. Okegbile, S.D., Gao, H., Talabi, O., Cai, J., Yi, C., Niyato, D., Shen, X.: FLES: a federated learning-enhanced semantic communication framework for mobile AIGC-driven human digital twins. IEEE Netw. (2025)
85. Ouedraogo, E.B., Hawbani, A., Wang, X., Liu, Z., Zhao, L., Al-Qaness, M.A.A., Alsamhi, S.H.: Digital twin data management: a comprehensive review. In: IEEE Transactions on Big Dat,a pp. 1–20 (2025)
86. Pamies Porras, M.J., Bertuch, T., Loecker, C., Adams, R., Wunderlich, R., Heinen, S.: An AESA antenna comprising an RF feeding network with strongly coupled antenna ports. IEEE Trans. Antennas Propag. **63**(1), 182–194 (2015)
87. Paulsson, K., Hubner, M., Becker, J.: Cost-and power optimized FPGA based system integration: methodologies and integration of a low-power capacity-based measurement application on Xilinx FPGAs. In: 2008 Design, Automation and Test in Europe, pp. 50–55 (2008)
88. Phan, H.A., Nguyen, P.V., Thi Khuat, T.H., Dang Van, H., Quoc Tran, D.H., Dang, B.L., Bui, T.T., Thi Thanh, V.N., Duc, T.C.: A sensor fusion approach for improving implementation speed and accuracy of rtab-map algorithm based indoor 3D mapping. In: International Joint Conference on Computer Science and Software Engineering, pp. 219–224 (2023)
89. Qian, Z., Chen, R., Yi, C., Zhai, X., Chen, B.: Collision avoidance control for autonomous driving with multiple dynamic obstacles in IoV: a prediction-enhanced APF-based approach. IEEE Internet Things J. (2025)
90. Qin, L., Lu, H., Chen, Y., Chong, B., Guo, F.: Joint transmission and resource optimization in noma-assisted IOVT with mobile edge computing. IEEE Trans. Veh. Technol. **73**(7), 9984–9999 (2024)
91. Qiu, H., Zhu, K., Luong, N.C., Yi, C., Niyato, D., Kim, D.I.: Applications of auction and mechanism design in edge computing: a survey. IEEE Trans. Cogn. Commun. Netw. **8**(2), 1034–1058 (2022)
92. Qu, S., Cui, J., Cao, Z., Qiao, Y., Men, X., Fu, Y.: Position estimation method for small drones based on the fusion of multisource, multimodal data and digital twins. Electronics (2024)
93. Ramzan, Z., Asif, H.M.S., Yousuf, I., Shahbaz, M.: A multimodal data fusion and deep neural networks based technique for tea yield estimation in Pakistan using satellite imagery. IEEE Access **11**, 42578–42594 (2023)
94. Raskar, R., van Baar, J., Beardsley, P., Willwacher, T., Rao, S., Forlines, C.: iLamps: geometrically aware and self-configuring projectors. In: Proceedings of ACM SIGGRAPH, pp. 809–818. ACM (2003)
95. Rathor, V.S., Garg, B., Sharma, G.K.: An energy-efficient trusted FSM design technique to thwart fault injection and Trojan attacks. In: 2018 31st International Conference on VLSI Design and 2018 17th International Conference on Embedded Systems (VLSID), pp. 73–78 (2018)
96. Razzaq, S., Shah, B., Iqbal, F., Ilyas, M., Maqbool, F., Rocha, A.: Deepclassrooms: a deep learning based digital twin framework for on-campus class rooms. Neural Comput. Appl. **35**(11), 8017–8026 (2023)
97. Rhee, J.M., Altaha, I.R.A., Hoang, D.N.M., Kim, D.H., Yang, J.S., Park, S.Y.: Seamless can: Can bus plus high-availability seamless redundancy. In: 2021 International Conference on Information and Communication Technology Convergence, pp. 1284–1286 (2021)

98. Romare, P., Morel, V., Karegar, F., Fischer-Hübner, S.: Tapping into privacy: a study of user preferences and concerns on trigger-action platforms. In: Proceedings of IEEE PST, pp. 1–12 (2023)
99. S Križanić, N.V.: Development of a process digital twin in Camunda modeler. TEM J. **13**(4), 3315–3323 (2024)
100. Sai, S., Hassija, V., Chamola, V., Guizani, M.: Federated learning and NFT-based privacy-preserving medical-data-sharing scheme for intelligent diagnosis in smart healthcare. IEEE Internet Things J. **11**(4), 5568–5577 (2024)
101. Shang, S., Yi, C., Zhang, T., Chen, R., Cai, J.: Latency-energy aware dynamic application placement for edge computing: a vacation queue based optimization approach. IEEE Trans. Netw. Sci. Eng. **11**(2), 2249–2263 (2024)
102. Shi, Y., Wang, J., Yi, C., Wang, R., Chen, B.: Proactive application deployment for MEC: an adaptive optimization based on imperfect multi-dimensional prediction. IEEE Trans. Veh, Technol (2025)
103. Shi, Y., Yang, Y., Yi, C., Chen, B., Cai, J.: Towards online reliability-enhanced microservice deployment with layer sharing in edge computing. IEEE Internet Things J. **11**(13), 23370–23383 (2024)
104. Shi, Y., Yi, C., Chen, B., Yang, C., Zhai, X., Cai, J.: Closed-loop control of edge-cloud collaboration enabled IIoT: an online optimization approach. In: Proceedings of IEEE ICC, pp. 5682–5687 (2022)
105. Shi, Y., Yi, C., Chen, B., Yang, C., Zhu, K., Cai, J.: Joint online optimization of data sampling rate and preprocessing mode for edge-cloud collaboration enabled industrial IoT. IEEE Internet Things J. **9**(17), 16402–16417 (2022)
106. Shi, Y., Yi, C., Wang, R., Wu, Q., Chen, B., Cai, J.: Service migration or task rerouting: a two-timescale online resource optimization for MEC. IEEE Trans. Wirel. Commun. **23**(2), 1503–1519 (2024)
107. Shin, H., Oh, S., Lee, J., Chung, G., Kim, J.: Designing data pipeline for network data management in digital twin network environment. In: Proceedings of IEEE ICAIIC, pp. 750–753 (2024)
108. Sudharani, S., Immanuel, D.G.: Control and analysis of UPQC for power quality improvement in renewable energy sources: advancing wireless power transfer and energy harvesting technologies. In: 2024 4th International Conference on Technological Advancements in Computational Sciences, pp. 383–388 (2024)
109. Sun, D., Sun, C., Ng, A., Cai, W.: Real differences between OT and CRDT in correctness and complexity for consistency maintenance in co-editors. Proc. ACM Hum.-Comput. Interact. **4**(CSCW1), 1–30 (2020)
110. Sun, K., Xu, H., Zhang, X., Hua, K., Li, J.: Time-sensitive local differential privacy-based federated learning for vehicular digital twin networks. In: International Symposium on Networks, Computers and Communications (2024)
111. Sun, W., Li, Y., Li, F., Liu, G., Rao, P.: Complex cloud-sea background simulation for space-based infrared payload digital twin. IEEE J. Sel. Topics Appl. Earth Observ. **18**, 3025–3042 (2025)
112. Taipalus, T.: Teaching tip: a notation for planning SQL queries. J. Inf. Syst. Educ. **30** (2019)
113. Team, D.: Building a real-time dashboard with Streamlit and apache Kafka. Dev3lop Blog (2025)
114. Team, E.: MQTT trends for 2025 and beyond: powering the future of AI and IoT. EMQX Blog (2025)
115. Team, P.: Matter, thread, and Zigbee: shaping the future of smart homes and IoT ecosystems. Promwad Blog (2025)
116. Tse, Y., Chen, V., Ng, W., et al.: A paradigm for collaborative 3D editing via list conflict-free replicated data types. In: Proceedings of ACM Symposium on Virtual Reality Software and Technology, pp. 1–11 (2022)
117. Turco, L., Zhao, J., Xu, Y., Tsourdos, A.: A study on co-simulation digital twin with Matlab and AIRSIM for future advanced air mobility. In: 2024 IEEE Aerospace Conference, pp. 1–18. IEEE (2024)

118. Ueno, Y., Tsukahara, A., Miyaho, N.: Next generation connectionless IP router architecture with switching delay for URLLC services. In: 2023 IEEE 24th Int. Conf. High Perform. Switching Routing, pp. 1–6 (2023)
119. Vakunov, A., Chang, C.L., Zhang, F., Sung, G., Grundmann, M., Bazarevsky, V.: Mediapipe hands: on-device real-time hand tracking. In: Workshop on Computer Vision for AR/VR, vol. 2, p. 5 (2020)
120. Wang, R., Wang, H., Zhu, K., Yi, C., Wang, P., Niyato, D.: Mobile charging services for internet of electric vehicles: concepts, scenarios and challenges. IEEE Veh. Technol. Mag. **18**(3), 110–119 (2023)
121. Wang, R., Wu, R., Liu, L., Yi, C., Zhu, K., Wang, P., Niyato, D.: Joint energy and computation workload management for geo-distributed data centers. IEEE Trans. Green Commun, Netw (2025)
122. Wang, R., Yu, X., Wu, Q., Yi, C., Wang, P., Niyato, D.: Efficient deployment of partial parallelized service function chains in CPU+DPU-based heterogeneous NFV platforms. IEEE Trans. Mob. Comput. **23**(10), 9090–9107 (2024)
123. Wang, X., Ashikhmin, A., Wang, X.: Wirelessly powered cell-free IoT: analysis and optimization. IEEE Internet Things J. **7**(9), 8384–8396 (2020)
124. Wen, J., Gabrys, B., Musial, K.: Evolutionary digital twin-oriented complex networked systems driven by node features and the mutation of feature preferences. PLoS One **19**(5), e0303571 (2024)
125. Wen, J., Nie, J., Zhong, Y., Yi, C., Li, X., Jin, J., Zhang, Y., Niyato, D.: Diffusion model-based incentive mechanism with prospect theory for edge AIGC services in 6G IoT. IEEE Internet Things J. **11**(21), 34187–34201 (2024)
126. Wu, J., Chen, R., Yi, C.: A DRL-based deviation-aware federated digital twin construction over wireless edge network. In: International Conference on Wireless Artificial Intelligent Computing Systems and Applications (WASA), pp. 196–208. Springer (2025)
127. Wu, J., Zhu, J., Zhang, J., Dang, P., Li, W., Guo, Y., Fu, L., Lai, J., You, J., Xie, Y., et al.: A dynamic holographic modelling method of digital twin scenes for bridge construction. Int. J. Digit. Earth **16**(1), 2404–2425 (2023)
128. Wu, K., Chen, J., Chen, L., Liu, Z., Yi, C., Xu, S., Wang, J., Kang, J.: QoE-aware joint visual and haptic signal transmission with adaptive data compression for immersive interactions in human digital twin. IEEE Trans. Netw. Service Manag. **22**(3), 2780–2794 (2025)
129. Wu, K., Chen, J., Yi, C., Liu, Z., Lu, X., Wang, J., Cai, J.: Joint visual and haptic signal transmission for immersive interactions in human digital twin. In: International Conference on Ubiquitous Communication (Ucom), pp. 465–470. IEEE (2024)
130. Wu, Q., Wang, R., Duan, X., Yi, C., Wang, P.: A lean networking framework (LeaNet): potential technical space and approaches for latency sensitive services. IEEE Netw. **38**(1), 280–287 (2024)
131. Wähner, K.: Iot architectures for digital twin with apache Kafka. DZone (2019)
132. Xiang, H., Wu, K., Chen, J., Cai, J., Niyato, D., Shen, X.: Realizing immersive communications in human digital twin by edge computing empowered tactile internet: visions and case study. IEEE Netw. **39**(4), 271–279 (2025)
133. Xiang, H., Zhang, T., Chen, J., Yi, C.: Joint optimization of feedback signal transmission and reconstruction in tactile internet. In: Proceedings of IEEE GLOBECOM, Taipei, Taiwan (2025)
134. Xiang, H., Zhang, T., Yi, C.: An optimization of feedback signal transmission and reconstruction for tactile internet: a multi-objective Bayesian approach. In: IEEE Transactions on Network Science and Engineering, pp. 1–18 (2025)
135. Xiao, S., Zhang, H., Yao, M., Cui, C., Li, J., Huang, C., Wang, C.X.: Demo: a novel 3D environment-aware digital twin online channel modeling platform. In: Proceedings of IEEE ICCC, pp. 297–298 (2024)
136. Xiong, P., Kang, Y., Liu, D., Xiao, F., Jiang, K., Yang, B.: Construction and commissioning of new generation renewable energy power system based on multi-port energy router. In: 2023 26th International Conference on Electrical Machines and Systems, pp. 749–754 (2023)

137. Yang, Y., Shi, Y., Chen, R., Yi, C., Kang, J.: Online optimization of edge empowered human digital twin deployment and task offloading. In: Proceedings of IEEE ICCC, pp. 856–861 (2024)
138. Yang, Y., Shi, Y., Yi, C., Cai, J., Kang, J., Niyato, D., Shen, X.: Dynamic human digital twin deployment at the edge for task execution: a two-timescale accuracy-aware online optimization. IEEE Trans. Mob. Comput. **23**(12), 12262–12279 (2024)
139. Yi, C., Alfa, A.S., Cai, J.: An incentive-compatible mechanism for transmission scheduling of delay-sensitive medical packets in e-health networks. IEEE Trans. Mob. Comput. **15**(10), 2424–2436 (2016)
140. Yi, C., Cai, J.: Multi-item spectrum auction for recall-based cognitive radio networks with multiple heterogeneous secondary users. IEEE Trans. Veh. Technol. **64**(2), 781–792 (2014)
141. Yi, C., Cai, J.: A priority-aware truthful mechanism for supporting multi-class delay-sensitive medical packet transmissions in e-health networks. IEEE Trans. Mob. Comput. **16**(9), 2422–2435 (2017)
142. Yi, C., Cai, J.: Ascending-price progressive spectrum auction for cognitive radio networks with power-constrained multi-radio secondary users. IEEE Trans. Veh. Technol. **67**(1), 781–794 (2018)
143. Yi, C., Cai, J.: Transmission management of delay-sensitive medical packets in beyond wireless body area networks: a queueing game approach. IEEE Trans. Mobile Comput. **17**(9), 2209–2222 (2018)
144. Yi, C., Cai, J.: A truthful mechanism for delay-dependent prioritized medical packet transmission scheduling. In: Proceedings of IEEE GLOBECOM, pp. 1–6 (2018)
145. Yi, C., Cai, J.: Delay-dependent priority-aware transmission scheduling for e-health networks: a mechanism design approach. IEEE Trans. Veh. Technol. **68**(7), 6997–7010 (2019)
146. Yi, C., Cai, J.: A queueing game approach for fog computing with strategic computing speed control. In: Proceedings of IEEE GLOBECOM, pp. 1–6 (2019)
147. Yi, C., Cai, J.: A truthful mechanism for scheduling delay-constrained wireless transmissions in IoT-based healthcare networks. IEEE Trans. Wirel. Commun. **18**(2), 912–925 (2019)
148. Yi, C., Cai, J., Su, Z.: A multi-user mobile computation offloading and transmission scheduling mechanism for delay-sensitive applications. IEEE Trans. Mob. Comput. **19**(1), 29–43 (2020)
149. Yi, C., Cai, J., Zhang, G.: Online spectrum auction in cognitive radio networks with uncertain activities of primary users. In: Proceedings of IEEE ICC, pp. 7576–7581 (2015)
150. Yi, C., Cai, J., Zhang, G.: Spectrum auction for differential secondary wireless service provisioning with time-dependent valuation information. IEEE Trans. Wirel. Commun. **16**(1), 206–220 (2017)
151. Yi, C., Cai, J., Zhu, K., Wang, R.: A queueing game based management framework for fog computing with strategic computing speed control. IEEE Trans. Mob. Comput. **21**(5), 1537–1551 (2022)
152. Yi, C., Ferens, K.: An energy efficient and minimum latency routing protocol for multihop WSNs. In: International Conferences on Wireless Networks (ICWN) (2013)
153. Yi, C., Huang, S., Cai, J.: Joint resource allocation for device-to-device communication assisted fog computing. IEEE Trans. Mob. Comput. **20**(3), 1076–1091 (2021)
154. Yi, C., Zhao, Z., Cai, J., Faria, R.L., Zhang, G.: Priority-aware pricing-based capacity sharing for beyond-wireless body area networks. Comput. Netw. **98**, 29–43 (2016)
155. Ying, C., Zhao, Z., Yi, C., Shi, Y., Cai, J.: An AoTI-driven joint sampling frequency and access selection optimization for industrial wireless sensor networks. IEEE Trans. Veh. Technol. **72**(9), 12311–12325 (2023)
156. Yu, B., Fang, L., Luo, L., Hu, X., Shen, C.: A study on service-oriented digital twin modeling methods for weaving workshops. Machines **12**(8), 542 (2024)
157. Yu, J., Yi, C., Zhang, T., Zhu, F., Cai, J.: A time utility function driven scheduling scheme for managing mixed-criticality traffic in TSN. In: International Conference on Wireless Algorithms, Systems, and Applications, pp. 20–28. Springer (2022)
158. Yu, Q., Zhang, T., Yi, C.: Consistent low-latency scheduling for microsecond-scale tasks in data centers. In: International Conference on Wireless Artificial Intelligent Computing Systems and Applications (WASA), pp. 100–113. Springer (2024)

159. Yu, X., Yang, Y., Chen, R., Yi, C.: Online optimization of edge vehicle digital twin migration with adaptive mobility prediction. In: Proceedings of IEEE VTC, Chengdu, China (2025)
160. Yuan, Y., Yi, C., Chen, B., Shi, Y., Cai, J.: A computation offloading game for jointly managing local pre-processing time-length and priority selection in edge computing. IEEE Trans. Veh. Technol. **71**(9), 9868–9883 (2022)
161. Zakharov, L., Derksen, L.: Hardware and software infrastructure of digital twin technology. KnE Eng. **5**(3), 29–35 (2020)
162. Zhang, S., Sun, Z., Chu, N., Wang, L., Li, L., Cai, C., Mohammad-Djafari, A.: Thermal fault detection of high-speed direct-driven blower components using thermal-visible image fusion and semantic segmentation. IEEE Sens. J. **25**(12), 21151–21161 (2025)
163. Zhang, Z., Yang, Z., Chen, M., Huang, C.: Energy efficient semantic communication over wireless networks with rate splitting. IEEE J. Sel. Areas Commun. **41**, 1484–1495 (2023)
164. Zhang, Z., Zhao, Y., Li, H., Lin, C., Liu, J.: DVFO: learning-based DVFs for energy-efficient edge-cloud collaborative inference. IEEE Trans. Mob. Comput. **23**(10), 9042–9059 (2024)
165. Zhao, S., Gong, S., Gu, B., Li, L., Lyu, B., Hoang, D.T., Yi, C.: Exploiting noma transmissions in multi-UAV-assisted wireless networks: from aerial-RIS to mode-switching UAVs. IEEE Trans. Wirel. Commun. **24**(3), 2530–2544 (2025)
166. Zhao, Y., Jian, M.: Applications and challenges of reconfigurable intelligent surface for 6G networks (original published in Chinese). Comput. Res. Repository (2021). arXiv:2108.13164
167. Zheng, H., Zhu, H., Wu, H., Yi, C.: Hierarchical reinforcement learning based collaborative control for industrial unmanned systems with semi-closed-loop feedbacks. In: Proceedings of IEEE ICUS, Changzhou, China (2025)
168. Zheng, H., Zhu, H., Wu, H., Yi, C., Zhu, K., Dai, X.: A DRL-based edge intelligent servo control with semi-closed-loop feedbacks in industrial IoT. In: International Conference on Wireless Artificial Intelligent Computing Systems and Applications (WASA), pp. 413–425. Springer (2024)
169. Zhou, H., Chen, R., Yi, C., Li, J., Cai, J.: A three-party repeated coalition formation game for PLS in wireless communications with IRSS. In: Proceedings of IEEE WCNC, pp. 1–6 (2024)
170. Zhou, H., Chen, R., Yi, C., Zhang, J., Kang, J., Cai, J., Guizani, M.: A repeated coalition formation game for physical layer security aware wireless communications with third-party intelligent reflecting surfaces. IEEE Trans. Wirel, Commun (2025)
171. Zhu, J., Wang, J., Huang, Y., He, S., You, X., Yang, L.: On optimal power allocation for downlink non-orthogonal multiple access systems. IEEE J. Sel. Areas Commun. **35**(12), 2744–2757 (2017)
172. Zhu, W., Chen, R., Yi, C., Cai, J.: Edge-assisted video transmission with adaptive key frame selection: a hierarchical DRL approach. In: Proceedings of IEEE Biennial Symposium on Communications, pp. 89–94. IEEE (2023)
173. Zong, Y., Dai, X., Wei, Z., Zou, M., Guo, W., Gao, Z.: Robust time synchronization for industrial Internet of Things by $h\infty$ output feedback control. IEEE Internet Things J. **10**(3), 2021–2030 (2023)
174. Zotter, F., Frank, M.: Ambisonics: a practical 3D audio theory for recording, studio production, sound reinforcement, and virtual reality. Appl. Sci. **9**(8), 1509 (2019)

Chapter 3
Federated Digital Twin Construction Under Dynamic Evolution

3.1 Introduction

Digital twin (DT) has emerged as a promising technology to create high-fidelity and hyper-realistic virtual counterparts of physical entities, revolutionizing the interaction between physical and digital worlds [63]. With predictive insights and real-time simulations, DT can drive breakthroughs across industries, such as remote healthcare, traffic management, autonomous driving and infrastructure planning [10, 11, 35]. To unleash the full potential, the DT construction, which refers to constructing a living and self-evolving visualization model empowered by artificial intelligence (AI) algorithms and deep neural networks [27], is obviously crucial. This requires not only the collection of real-time sensing data acquired from the ever-changing environment, but also the continuous state updates, to guarantee the synchronization between the DT's virtual model and its corresponding physical entity [39, 56].

Although the DT construction via distributed sensing has been discussed in the existing work [11, 11, 27, 56], all of them were restricted to the centralized framework, where the feature data is collected by deployed sensors, and directly transmitted to a central cloud for processing (including fine-grained model docking, rendering and visualization [1, 32]). However, as the status of physical entity always varies dynamically over the time, such a centralized framework has to be maintained by continuously uploading all data to the central cloud and updating the entire DT model [28], which may incur significant communication and computation overheads. Moreover, the prevalence of data silos in wide-range physical environments may also impede the real-time data collection and utilization [8, 11, 46], which ultimately degrades the quality of the centrally constructed DT model. To overcome these bottlenecks, we introduce, for the first time, a novel federated DT construction framework under edge-cloud collaboration. In this framework, a DT model to be built at the central cloud can be regarded as a global one being split into and integrating from multiple functional components, i.e., partial-DTs, created at edge servers (ESs). Specifically, (i) the central cloud, which is responsible for constructing the

C. Yi et al., *Self-Evolving Digital Twin Over Wireless Networks*, Wireless Networks,
https://doi.org/10.1007/978-3-032-08120-9_3

DT model, first splits the whole model into multiple partial-DTs; (ii) each partial-DT is then assigned to an ES for creation using feature data provided by its associated sensors; and (iii) once all partial-DTs are created, they are forwarded to the central cloud for global integration, i.e., connecting and harmonizing these partial-DTs into a complete global DT model. In practice, this federated DT construction framework can be applied to various practical scenes, such as intelligent transportation and smart factory, where the local server at each intersection or factory creates partial-DTs for road segments or assembly lines using surveillance data collected by traffic or industrial cameras, and then transmits its partial-DT to the headquarter [28, 42, 54, 67]. By this way, the DT construction is performed with the help of partial-DTs' creations and evolutions in distributed and parallel structures, so that high cost-efficiency and DT model quality can be achieved.

Essentially, the DT model is a complex integration of not only AI models, but also visual models, geometric models, rule models, data analytic models, etc. [43, 54]. While the federated model construction, typically the federated learning, has been extensively studied, the federated DT construction possesses unique features, making it fundamentally different, and thus is worthy to be carefully explored. On one hand, unlike the homogeneous local models trained in the federated learning, which all deal with the same task [40, 50, 55, 58, 62], partial-DTs in the federated DT construction are heterogeneous, each with distinct model parameters and functions, trained by potentially various sources of feature data. For instance, in the transportation DT system, partial-DTs for vehicles, pedestrians and road scenes can be diverse due to different monitoring areas or sensing equipments, but should be simultaneously created and later integrated as a whole [20, 66]. On the other hand, unlike the federated learning that commonly focuses on training a global model until convergence in a single time frame, the federated DT construction has to be dynamically conducted frame-by-frame in accordance with the DT evolution. For example, the transportation DT should be timely updated to synchronize the real-world road traffics, so that its construction is adaptive rather than static [28, 56]. For exploiting the benefit of the federated DT construction and facilitating its implementation under edge-cloud collaboration, it is required to optimize partial-DT assignments from the cloud to ESs, ES-sensor associations for partial-DT creation, and as well as computation and communication resource allocations for global DT integration. Despite the significance of this issue, it is very challenging for the following reasons.

1. Evidently, under edge-cloud collaboration, DT model splitting, partial-DT assignment, ES-sensor association, partial-DT training and global DT integration are sequential but tightly coupled, meaning that all corresponding resource allocations have to be jointly optimized. Moreover, unlike the federated learning where each local model is statically weighted by a predetermined coefficient, partial-DTs created at ESs may contribute differently to the integration of the global DT model at the cloud [7, 23, 66], having different importance weights. Caused by uncertain DT evolutions, such importance weights may be time-varying. This together with unpredictable wireless networking conditions and sensing capabilities, make the optimization uncertain and dynamic, motivating us to design an

advanced online algorithm with strong exploration capability for maximizing the system performance in long runs.

2. In practice, partial-DTs, as multiple functional components of a global DT model, may exhibit certain shared characteristics, such as the same rendering material and kernel functions, implying that the data required for their creations are partly identical [23, 32]. As a result, in our proposed federated DT construction framework, each sensor can simultaneously associate with multiple ESs, participating in different partial-DT creations. This leads to a complicated interrelationship between ESs and sensors, i.e., overlapped ES-sensor associations, necessitating us to integrate the online optimization algorithm by an extension of the analysis and solution in conventional exclusively independent coalitions [9, 38] to overlapping coalitions.

In this chapter, to fill the gap in the literature, we propose a novel federated DT construction framework with the support of edge-cloud collaboration. We aim to maximize the global DT's model quality while minimizing all induced costs, including energy consumption and configuration costs. By taking into account uncertain DT evolutions and resulted impacts, we formulate an online optimization problem for dynamically determining partial-DT assignments, ES-sensor associations, and computation and communication resource allocations. To circumvent the difficulty in solving this complicated problem, we first transform it into a two-layer hierarchical game. In the upper layer, a two-sided matching subgame is formulated for the cloud to optimize partial-DT assignments. In the lower layer, an overlapping coalition formation game is formulated for ESs to optimize ES-sensor associations and resource allocations. After proving an existence of the equilibria within these games, we apply the Gale-Shapley (GS) algorithm and particularly develop a switch rules-based overlapping coalition formation (SOCF) algorithm to find the short-term equilibria of upper-layer and lower-layer subgames, respectively. Then, we design a deep reinforcement learning (DRL) based solution integrated with stable matching and overlapping coalition formation, called DMO, addressing the dynamic and multi-dimensional decision making problem. The DMO approach captures the time interdependence, and connects the short-term equilibria to a long-term equilibrium of the hierarchical game, providing an efficient solution to the original online optimization problem.

The main contributions of this chapter are summarized in the following:

- To the best of our knowledge, we are the first to introduce the federated DT construction via distributed sensing under edge-cloud collaboration. We formulate an online optimization problem that dynamically optimizes partial-DT assignments, ES-sensor associations, and computation and communication resource allocations to maximize the long-term system performance of the considered framework.
- We propose a novel game-theoretical solution that first transforms the online optimization problem into a two-layer hierarchical game. Then, based on the equilibrium analyses, we leverage the GS algorithm and develop an SOCF algorithm to efficiently obtain the short-term equilibria of upper and lower subgames. On

top of this, we further design the DMO approach to well accommodate dynamic settings.

- We conduct extensive simulations to evaluate the performance of the proposed federated DT construction framework along with the game-theoretic optimization solution. Results show the effectiveness of the framework and the superiority of the solution in terms of increasing the gain while reducing the cost in the DT construction compared to counterparts.

The rest of this chapter is organized as follows: Sect. 3.2 reviews the related work and highlights the novelties of this chapter. Section 3.3 presents the system model of the federated DT construction framework and formulates the corresponding optimization problem. In Sect. 3.4, a game-theoretic solution approach with two-sided matching and overlapping coalitions is proposed and analyzed. Simulation results are given in Sect. 3.5, followed by the conclusion in Sect. 3.6.

3.2 Related Work

As a key aspect to the promotion of DT applications, the DT construction has drawn dramatically increasing attentions from both academia and industry. For example, Microsoft released the Azure DT platform [34] which allowed users to create various DT models on the cloud. A startup company called Carbontribe, offered an AI-driven DT prototype [6] that can accurately predict the greenhouse gas emissions using satellite and biodiversity data. Besides, the DT construction over networks, enabled by edge computing, distributed sensing, etc. has also be recently studied. For instance, Yang et al. [56] explored a vehicular DT construction framework, in which the traffic DT was built at the central cloud and dynamically synchronized with the moving vehicles, assisted by the relay of roadside units. Li et al. [27] proposed a continual learning-driven DT construction framework to build and update the DT on multiple ESs using the data collected from Internet-of-Things devices, aiming to maximize the long-term model accuracy. Chen et al. [11], Yang et al. [57], Yu et al. [61] developed a personalized DT construction framework, where DTs with both generic and customized parts were dynamically deployed on local ESs under the control of a remote cloud, utilizing the collected feature data from nearby sensors. However, all these work focused on the centralized DT construction framework, ignoring the potential of constructing the DT in a federated manner.

Some researchers have investigated the federated model construction, particularly the federated learning and its variants. Name a few, Zhang et al. [62] studied the intelligent reflecting surfaces-assisted federated learning, where identical models were trained on local clients, targeting to minimize the energy consumption till the convergence of model accuracy. Pham et al. [40] developed an air-ground federated learning structure, where multiple ground clients were hired to first train sub-models for the same task, and the global model was then aggregated at a drone server. Further taking into account the system dynamics, Sun et al. [50] devised a Stackelberg game-empowered federated learning framework, where clients were asked to perform the same task and were dynamically selected via an incentive mechanism, assisted by a DT of the whole air-ground network. However, the solutions in all these work can hardly be employed for the DT construction because unlike general models, the DT should be dynamically evolving with the ever-changing physical environments and is composed by heterogeneous partial-DTs, each of which contributes differently to the global DT.

Some preliminary efforts have been devoted to studying the online optimization of task splitting and assignments, client associations, and resource allocations, that may possibly applied for the federated DT construction. For example, Kobayashi et al. [25] formulated a dynamic combinatorial optimization problem for jointly optimize the channel assignment and base station association for downlink-uplink decoupling in wireless networks. Ng et al. [38] proposed a reputation-aware hedonic coalition formation game approach to divide the clients into multiple non-overlapping groups for associating with an intermediate aggregator within a hierarchical federated learning framework. Xu et al. [55] designed an online learning algorithm to dynamically and jointly select a non-overlapping subset of clients associated with one aggregator, which was then assigned to perform one federated learning request, aiming to maximize the long-term energy efficiency. Nevertheless, all these work assumed that tasks were independent with each other, so that client associations were exclusive without any overlaps. This does not fit the federated DT construction framework under edge-cloud collaboration, where each sensor may be associated with multiple ESs simultaneously, participating in various partial-DT creations.

In summary, this work differs from the existing literature in the following aspects: (i) we introduce a federated DT construction framework supported by edge-cloud collaboration to achieve higher model quality and lower costs in a distributed manner, rather than the traditional centralized ones; (ii) we consider the uncertain DT evolution, along with the heterogeneity of partial-DTs, and formulate an online optimization problem to jointly and dynamically determine partial-DT assignments, ES-sensor associations and resource allocations; (iii) we study overlapping coalitions in ES-sensor associations for partial-DT creation, instead of the conventional exclusively independent coalitions, and make them accommodate the dynamic settings.

3.3 System Model and Problem Description

In this section, we first present a system overview on how a DT model is constructed via distributed sensing and evolved in a federated manner. Then, we describe the partial-DT creation on local ESs and global DT integration on the central cloud. After that, aiming to create a high-quality DT model while reducing the costs from data sensing, partial-DT creation and global DT integration in long runs, we formulate an online optimization problem.

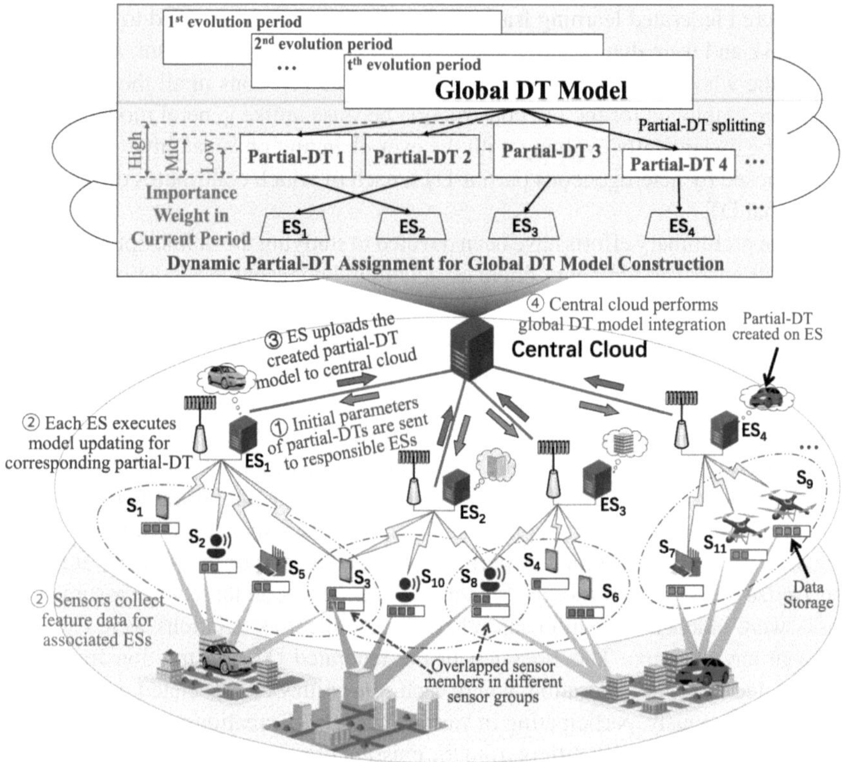

Fig. 3.1 An illustration of the federated DT construction by edge-cloud collaboration

3.3.1 System Overview

Consider a federated DT construction framework building upon an edge-cloud collaborative system, as shown in Fig. 3.1, consisting of a central cloud CS, a set of geographically dispersed local ESs $\mathcal{B}$ with $|\mathcal{B}| = B$, and a set of randomly scattered sensors $\mathcal{N}$ with $|\mathcal{N}| = N$. In this framework, the DT model is seen as a global one, which can be split into multiple functional components [23, 32, 66], i.e., partial-DTs, which are simultaneously created on ESs. These partial-DTs are essentially different partial digital replicas of the physical entity, and can be obtained by using the featured data collected from selected groups of associated sensors [24, 28, 32]. Then, integrating all partial-DTs from ESs to the central cloud, a global DT model can be constructed. Following the DT configurations [27, 50], we can define that

$$\text{Global-DT} = \big\langle \{\text{Partial-DT } c\}_{\forall c \in C} \big\rangle, \tag{3.1}$$

where C denotes the set of all partial-DTs of the global DT model, $c \in C$ indicates one specific partial-DT, and $\langle \cdot \rangle$ represents the integration of all partial-DTs.

Although each partial-DT $c \in C$ is created in a distributed manner on a specific ES, the central cloud has to first inform each ES by the initial values of core parameters in partial-DT c (e.g., model structures, rendering requirements, configuration information [23, 24]), performing the DT model splitting and partial-DT assignments. Then, sensors are divided into multiple groups, each group is associated with one ES for feature data collection. Here, since different ESs may require the same data for creating partial-DTs with shared characteristics, we do not restrict one sensor to be associated with at most one ES, implying that each of them can be potentially in multiple groups, simultaneously helping different ESs' partial-DT creations [2, 45]. By utilizing the initial values informed by the cloud and the feature data acquired by associated sensors in the physical-virtual mapping process, each ES can then establish a partial-DT model, for global DT integration.

Note that DT model should be timely constructed and evolved to maintain a high quality in long runs because its original goal is to replicate realistically the physical world which always vary dynamically [13, 14]. Consequently, the proposed federated DT construction procedure should evolve over the time, in which all partial-DTs has to be continuously recreated and updated by time-increasing data collected from sensors in real time. To capture the system dynamics stem from long-term evolution nature of the DT, we employ a time-framed system, where each time frame $t \in \{1, 2, ..., T\}$ refers to the duration of one DT evolution period. Furthermore, in practice, partial-DTs may contribute differently to the integration of the global DT model, meaning that they have different importance weights [7], and these weights can also vary with DT evolutions. Thus, we introduce a dynamic importance weight index, defined as $\mathcal{I} = \{I_c(t)\}_{c \in C}$, where $I_c(t) \in [0, 1]$ is the importance weight of partial-DT $c \in C$ in time frame t, which can be revealed by the specific application requirements of each partial-DT.

Overall, we aim to optimize the long-term system performance of the federated DT construction framework, i.e., enhancing the quality of the global DT model while reducing the costs from data sensing, partial-DT creation and global DT integration in long term. Particularly, in each time frame t, we are required to dynamically determine (i) which partial-DT should be assigned to which ES for creation, so that partial-DTs with larger importance weights can always be handled by ESs with both more computation resources and more associated sensors in feature data collection; and (ii) which sensors should be grouped to associate with which ES for assisting its partial-DT creation, allowing that one sensor can be associated with multiple ESs, forming overlapping groups. For convenience, the main notations used in this paper are listed in Table 3.1.

3.3.2 Model Quality of Federated DT Construction

In the federated DT construction framework, the quality of the global DT model constructed at the central cloud within each time frame t, denoted by $A^{Global}(t)$, depends on the partial-DT assignments, time-varying importance weight of each

Table 3.1 Important notations in this chapter

Symbol	Explanation
$A_{c,b}(t)$	Model quality of partial-DT c created on ES b
A_c^{Req}	Required maximum model accuracy of partial-DT c
$A^{Global}(t)$	Quality of the global DT model
C^{Conf}	Potential configuration cost
$Co_b(t)$	All sensors associated to ES b
$\mathcal{C}, \mathcal{B}, \mathcal{N}$	Set of partial-DTs, ESs or sensors
$\mathcal{CP}(t)$	Coalition partition
$d_c(t)$	Amount of data collected for partial-DT c
$E^{Total}(t)$	Total energy consumption
$\Phi_t(i)$	Bidirectional matching
$\mathcal{G}^H$	Hierarchical game
$\mathcal{G}_t^{Up}, \mathcal{G}_t^{Low}$	Upper-layer and lower-layer subgames
$I_c(t)$	Importance weight of partial-DT c
P_n	Transmit power of sensor n
$\pi_b^R(t)$	Resource allocation for ES b
$R_{n,b}^T(t)$	Wireless transmission rate from sensor n to ES b
$T_b(t)$	Decision on training rounds of ES b
$\tau^{Total}(t)$	Total latency
$\mathcal{U}(t), \mathcal{U}_{Sys}$	Short-term and long-term system performance
$x_{c,b}(t)$	Decision on partial-DT assignments
$y_{b,n}(t)$	Decision on ES-sensor associations
$z_{b,n,w}(t)$	Decision on subcarrier allocations

partial-DT and achieved model quality of each partial-DT created by the responsible ES, which can be expressed as

$$A^{Global}(t) = \sum\nolimits_{c\in\mathcal{C}} \sum\nolimits_{b\in\mathcal{B}} x_{c,b}(t) I_c(t) A_{c,b}(t), \tag{3.2}$$

where $I_c(t)$ stands for the importance weight of each partial-DT $c \in \mathcal{C}$ in time frame t, $A_{c,b}(t)$ represents the model quality of partial-DT $c \in \mathcal{C}$ created by ES $b \in \mathcal{B}$ in time frame t, and $x_{c,b}(t) \in \{0, 1\}$ is the decision on partial-DT assignment, indicating whether partial-DT $c \in \mathcal{C}$ is assigned to be created on ES $b \in \mathcal{B}$ in time frame t. Note that we impose the following constraints

$$\sum\nolimits_{b\in\mathcal{B}} x_{c,b}(t) = 1, \forall c \in \mathcal{C}, \text{ and } \sum\nolimits_{c\in\mathcal{C}} x_{c,b}(t) \leq 1, \forall b \in \mathcal{B}, \tag{3.3}$$

meaning that all partial-DTs have to be created for guaranteeing the global DT model integrity, and one ES is allowed to handle the creation of at most one partial-DT.

To further measure $A_{c,b}(t)$, we extend the well-known definition of model accuracy in the federated learning [40, 55, 58, 62] as the performance of partial-DT creations. Namely, $A_{c,b}(t)$ is calculated based on both the amount of cumulatively collected feature data and the training rounds that ES $b \in \mathcal{B}$ determines to take for creating partial-DT $c \in C$ under iteratively model training using the feature data.[1] Mathematically, $A_{c,b}(t)$ can be written as

$$A_{c,b}(t) = x_{c,b}(t) \min \left\{ A_c^{Req}, \Gamma\left(\sum\nolimits_{i=1}^{t} d_c(i)\right) \cdot \left(1 - 2^{(-T_b(t)(2-L\delta)\delta\gamma)/2}\right) \right\}, \quad (3.4)$$

where A_c^{Req} stands for the desired maximum model quality of partial-DT $c \in C$, $\Gamma\left(\sum_{i=1}^{t} d_c(i)\right)\left(1 - 2^{(-T_b(t)(2-L\delta)\delta\gamma)/2}\right)$ indicates the model quality that each partial-DT $c \in C$ can achieve after training multiple rounds with the cumulatively collected data [55, 58]. Particularly, $\Gamma\left(\sum_{i=1}^{t} d_c(i)\right) = \left(\log_2\left(\frac{\sum_{i=1}^{t} d_c(i)}{t\beta} + 1\right)\right)^2$ represents the impact from the cumulatively collected feature data following [11, 51], expressed as a discount coefficient calculated by logarithmic normalization function with parameter β [36], and $\sum_{i=1}^{t} d_c(i)$ denotes the amount of cumulatively collected feature data for partial-DT $c \in C$ up to time frame t. $1 - 2^{(-T_b(t)(2-L\delta)\delta\gamma)/2}$ calculates an approximate model accuracy in terms of the logistic loss function [55, 58, 62], where $T_b(t)$ is the number of training rounds determined to be taken by ES $b \in \mathcal{B}$ in time frame t, L, γ and δ are pre-defined hyperparameters, indicating that the loss function in training the AI model for DT satisfies L-Lipschitz continuous and γ-strongly convex, with $\delta \in (0, 2/L)$ being the corresponding learning rate [58].

Observed from (3.4) that $A_{c,b}(t)$ increases with $T_b(t)$ but its marginal benefit decreases when $T_b(t)$ is considerably large. To be cost-efficient, $T_b(t)$ is naturally constrained as

$$T_b(t) \leq T_b^*(t), \forall b \in \mathcal{B}, \quad (3.5)$$

where $T_b^*(t)$ is the upper bound of $T_b(t)$, which can be calculated by letting $A_{c,b}(t) = A_c^{Req}$ in (3.4), i.e.,

$$T_b^*(t) = \sum\nolimits_{c \in C} x_{c,b}(t) \left\lceil \frac{2}{(L\delta - 2)\delta\gamma} \log_2\left(1 - \frac{A_c^{Req}}{\Gamma\left(\sum_{i=1}^{t} d_c(i)\right)}\right) \right\rceil. \quad (3.6)$$

Furthermore, considering that the ES-sensor association plays a vital role in feature data collection for partial-DT creations, the total amount of new data collected in each time frame t for partial-DT $c \in C$ is expressed as

[1] Note that, the model quality of partial-DT is approximated by a weighted model accuracy with respect to the data size, and it is formulated in this particular way for mathematical tractability, while our proposed solution can adapt to other customized model quality measurement tailored to specific DT applications.

$$d_c(t) = \sum\nolimits_{b\in\mathcal{B}} x_{c,b}(t) \sum\nolimits_{\forall n\in\mathcal{N}} y_{b,n}(t) d_{n,c}(t), \tag{3.7}$$

where $d_{n,c}(t)$ stands for the amount of feature data that each sensor n can obtain for partial-DT $c \in C$ in time frame t, and $y_{b,n}(t) \in \{0, 1\}$ is the decision on the ES-sensor association, denoting whether each sensor $n \in \mathcal{N}$ is determined to be associated with ES $b \in \mathcal{B}$ in time frame t. Recall that we allow each sensor to be associated with multiple ESs, while the total number of ESs that one sensor can be associated with should be restricted due to the hardware limitations of the multi-cast transmissions [17, 19], i.e.,

$$\sum\nolimits_{b\in\mathcal{B}} y_{b,n}(t) \le L_n, \forall n \in \mathcal{N}, \tag{3.8}$$

where L_n denotes the maximum number of ESs with which each sensor $n \in \mathcal{N}$ can be associated.

3.3.3 Latency and Energy Consumption in Federated DT Construction

The total latency of federated DT construction in each time frame t is the sum of (i) the latency in preparing all partial-DTs at ESs, which includes sensing data transmission latency $\tau_b^{Dtr}(t)$, historical data migration latency $\tau_b^{Back}(t)$, partial-DT creation latency $\tau_b^{Cre}(t)$ and partial-DT uploading latency $\tau_b^{Mtr}(t)$; and (ii) the global DT integration latency $\tau_{CS}^{I}(t)$ at the cloud, i.e.,

$$\tau^{Total}(t) = \max_{b\in\mathcal{B}}\{\tau_b^{Dtr}(t) + \tau_b^{Back}(t) + \tau_b^{Cre}(t) + \tau_b^{Mtr}(t)\}\tau_{CS}^{I}(t), \tag{3.9}$$

where $\max_{b\in\mathcal{B}}\{\cdot\}$ calculates the maximum latency among all partial-DTs on all ESs. Hence, the total energy consumption of federated DT construction in each time frame t is

$$E^{Total}(t) = \sum\nolimits_{b\in\mathcal{B}} \left(E_b^{Dtr}(t) + E_b^{Back}(t) + E_b^{Cre}(t) + E_b^{Mtr}(t)\right) + E_{CS}^{I}(t), \tag{3.10}$$

where $E_b^{Dtr}(t)$, $E_b^{Back}(t)$, $E_b^{Cre}(t)$ and $E_b^{Mtr}(t)$ respectively denote the energy consumption of sensing data transmission, historical data migration, partial-DT creation and partial-DT uploading in preparing each partial-DT $c \in C$ on its responsible ES $b \in \mathcal{B}$, and $E_{CS}^{I}(t)$ represents the energy consumption of global DT integration at the cloud, all in time frame t.

Specifically, $\tau_b^{Dtr}(t)$ and $E_b^{Dtr}(t)$ are the maximum latency and total energy consumption of all sensors associated with each ES $b \in \mathcal{B}$, respectively, in collecting data for partial-DT $c \in C$, i.e.,

$$\tau_b^{Dtr}(t) = \max_{n \in \mathcal{N}} \{x_{c,b}(t) y_{b,n}(t) d_{n,c}(t) / R_{n,b}^T(t)\}, \tag{3.11}$$

$$E_b^{Dtr}(t) = \sum\nolimits_{n \in \mathcal{N}} y_{b,n}(t) P_n x_{c,b}(t) d_{n,c}(t) / R_{n,b}^T(t), \tag{3.12}$$

where $x_{c,b}(t) y_{b,n}(t) d_{n,c}(t)$ indicates the amount of sensing data acquired by each sensor $n \in \mathcal{N}$ associated with ES $b \in \mathcal{B}$ assigned to create partial-DT $c \in C$, P_n denotes the transmit power of sensor $n \in \mathcal{N}$, and $R_{n,b}^T(t)$ stands for the wireless transmission rate from sensor $n \in \mathcal{N}$ to ES $b \in \mathcal{B}$ in time frame t.

Particularly, $R_{n,b}^T(t)$ can be calculated as

$$R_{n,b}^T(t) = y_{b,n}(t) z_{b,n,w}(t) W \log_2 \Bigg(1 + \frac{P_n H_{n,b,w}(t)}{\left(\sum_{b' \in \mathcal{B}} \sum_{n' \in \mathcal{N}} y_{b',n'}(t) z_{b',n',w}(t) P_{n'} H_{n',b,w}(t) + n_b^2\right)}\Bigg), \tag{3.13}$$

where W represents the bandwidth of each subcarrier $w \in \mathcal{W}$, $H_{n,b,w}(t)$ is the instantaneous channel gain from wireless sensor $n \in \mathcal{N}$ to ES $b \in \mathcal{B}$ on subcarrier $w \in \mathcal{W}$, $\sum_{b' \in \mathcal{B}} \sum_{n' \in \mathcal{N}} y_{b',n'}(t) z_{b',n',w}(t) P_{n'} H_{n',b,w}(t)$ indicates the co-channel interference at ES $b \in \mathcal{B}$, n_b^2 signifies the additive white Gaussian noise at ES $b \in \mathcal{B}$, and $z_{b,n,w}(t) \in \{0, 1\}$ is the decision on subcarrier allocation [19], denoting whether each subcarrier $w \in \mathcal{W}$ is allocated to sensor $n \in \mathcal{N}$ that is associated with ES $b \in \mathcal{B}$ in time frame t. Obviously, we should have

$$\sum\nolimits_{n \in \mathcal{N}} \sum\nolimits_{w \in \mathcal{W}} y_{b,n}(t) z_{b,n,w}(t) \le |\mathcal{W}|, \tag{3.14}$$

where $|\mathcal{W}|$ is the maximum number of subcarriers that one ES can provide to all its associated sensors.

Besides the newly collected sensing data, historical data is also indispensable in the partial-DT creation. For each partial-DT $c \in C$ created on ES $b \in \mathcal{B}$, this data should be dynamically migrated from the previously assigned ES $b' \in \mathcal{B}$ in time frame $t-1$ to ES b in time frame t [47, 48]. Thus, we have

$$\tau_b^{Back}(t) = \sum\nolimits_{b' \in \mathcal{B}} x_{c,b'}(t-1) \sum\nolimits_{i=1}^{t-1} d_c(i) / R_{b',b}^T, \tag{3.15}$$

$$E_b^{Back}(t) = \sum\nolimits_{b' \in \mathcal{B}} x_{c,b'}(t-1) P_{b'} \sum\nolimits_{i=1}^{t-1} d_c(i) / R_{b',b}^T, \tag{3.16}$$

where $\sum_{i=1}^{t-1} d_c(i)$ indicates the amount of cumulatively historical data up to time frame $t-1$, $R_{b',b}^T$ stands for the transmission rate from ES $b' \in \mathcal{B}$ to another ES $b \in \mathcal{B}$ via the fiber link [11], and $P_{b'}$ is the transmit power of ES b'.

In each time frame t, since creating the assigned partial-DT $c \in C$ on ES $b \in \mathcal{B}$ involves the model training with multiple rounds $T_b(t)$ for improving the model quality, the latency and energy consumption for such partial-DT creation can be respectively written as

$$\tau_b^{Cre}(t) = T_b(t) \sum\nolimits_{n \in \mathcal{N}} \big(y_{b,n}(t) x_{c,b}(t) d_{n,c}(t) \Upsilon_b\big)/F_b, \tag{3.17}$$
$$E_b^{Cre}(t) = \rho_b (F_b)^2 T_b(t) \Upsilon_b, \tag{3.18}$$

where Υ_b stands for the CPU cycles required for computing one byte of data on ES $b \in \mathcal{B}$ (measured by s/byte) [58], F_b is the CPU speed of ES $b \in \mathcal{B}$ (measured by cycles/s), and ρ_b denotes the effective switched capacitance of ES $b \in \mathcal{B}$.

After the partial-DT creation, each ES $b \in \mathcal{B}$ will upload its created partial-DT $c \in C$ to the cloud. The corresponding latency and energy consumption can be respectively expressed as

$$\tau_b^{Mtr}(t) = x_{c,b}(t) D_c \big/ R_{b,CS}^T, \tag{3.19}$$
$$E_b^{Mtr}(t) = x_{c,b}(t) P_b D_c \big/ R_{CS,b}^T, \tag{3.20}$$

where D_c is the model size of each partial-DT $c \in C$, and $R_{b,CS}^T$ denotes the transmission rate from ES $b \in \mathcal{B}$ to the cloud via the fiber link [11].

Furthermore, the latency for global DT integration in each time slot t is caused by processing all partial-DTs aggregated at the cloud, i.e.,

$$\tau_{CS}^I(t) = \sum\nolimits_{c \in C} \sum\nolimits_{b \in \mathcal{B}} x_{c,b}(t) D_c \Upsilon_{CS} \big/ F_{CS}, \tag{3.21}$$

where Υ_{CS} is the number of CPU cycles required for integrating a single byte data of partial-DT models, and F_{CS} is the CPU speed of the cloud. The energy consumption $E_{CS}^I(t)$ of global DT integration due to executing resource-intensive instructions over both CPU and GPU resources on the cloud (e.g., fine-grained model docking, rendering and visualization [1, 32]), following [55], $E_{CS}^I(t)$ can be calculated as

$$E_{CS}^I(t) = \tau_{CS}^I(t) \Big(\sum\nolimits_{c \in C} \sum\nolimits_{b \in \mathcal{B}} x_{c,b}(t) \frac{\zeta_{CS} H_{CS,c}}{\tau_{CS}^I(t)} \cdot P_{CS}^{Max} + P_{CS}^{Idle} + P_{CS}^{Leak} \Big), \tag{3.22}$$

where ζ_{CS} is the access rate of the cloud's single processing unit, $H_{CS,c}$ is the average number of instructions in integrating partial-DT $c \in C$, P_{CS}^{Max}, P_{CS}^{Idle} and P_{CS}^{Leak} are hardware parameters related to the cloud's processing units [55].

3.3.4 Problem Formulation

To evaluate the performance of the federated DT construction framework, for each single time frame t, we take the difference between (i) the gain of the global DT model quality $A^{Global}(t)$, and (ii) the cost of total energy consumption $E^{Total}(t)$, as

$$\mathcal{U}(t) = \xi A^{Global}(t) - \kappa E^{Total}(t), \tag{3.23}$$

where ξ and κ are weight coefficients.

Furthermore, when expanding $\mathcal{U}(t)$ over multiple time frames for the long-term optimization, partial-DT assignments among ESs can be dynamically adjusted, and triggered by this, ES-sensor associations may also be dynamically varied. This incurs an additional configuration cost between any two consecutive time frames, e.g., signaling overheads caused by notifying such changes [15, 64, 65]. Considering this, we define the long-term system performance of the federated DT construction framework as

$$\mathcal{U}_{Sys} = \lim_{T\to\infty} \frac{1}{T} \sum\nolimits_{t=0}^{T} \mathcal{U}(t) - C^{Conf} \mathcal{F}(t), \tag{3.24}$$

where C^{Conf} is a pre-known cost coefficient, and $\mathcal{F}(t) = \sum_{b\in\mathcal{B}} \sum_{n\in\mathcal{N}} y_{b,n}(t) \oplus y_{b,n}(t-1)$ indicates the total number of changes of ES-sensor associations in time frame t, with $\oplus$ being the exclusive OR operator.

Then, with the objective of maximizing $\mathcal{U}_{Sys}$, we formulate a long-term optimization problem to jointly and dynamically determine partial-DT assignments $x_{c,b}(t)$, ES-sensor associations $y_{b,n}(t)$, subcarrier allocations $z_{b,n,w}(t)$ and training rounds $T_b(t)$ in every time frame t, denoted as $\Pi_{Sys} = \big\{x_{c,b}(t), y_{b,n}(t), z_{b,n,w}(t), T_b(t)\big\}_{\forall t,c,b,w}$, which is given bellow.

$$[\mathcal{P}_1]: \max_{\Pi_{Sys}} \mathcal{U}_{Sys} \tag{3.25a}$$

$$s.t., \ (3.3),\ (3.8),\ (3.14),$$

$$\tau^{Total}(t) \le \delta_T, \tag{3.25b}$$

$$R^T_{n,b}(t) \ge R^T_{Min}, \forall b \in \mathcal{B}, \forall n \in \mathcal{N}, \tag{3.25c}$$

$$x_{c,b}(t) \in \{0, 1\}, \ \forall c \in \mathcal{C}, \forall b \in \mathcal{B}, \tag{3.25d}$$

$$y_{b,n}(t) \in \{0, 1\}, \ \forall b \in \mathcal{B}, \forall n \in \mathcal{N}, \tag{3.25e}$$

$$z_{b,n,w}(t) \in \{0, 1\}, \ \forall b \in \mathcal{B}, \forall n \in \mathcal{N}, \forall w \in \mathcal{W}, \tag{3.25f}$$

$$T_b(t) \le T^*_b(t), \forall b \in \mathcal{B}, \tag{3.25g}$$

where constraint (3.25b) enforces that the cloud should integrate all partial-DTs for model integrity within δ_T, which stands for the time duration of one DT evolution period; constraint (3.25c) ensures that $R^T_{n,b}(t)$ should not be smaller than a minimum value R^T_{Min} for stable communications between sensors and ESs, and constraints (3.25d)–(3.25g) are the defined ranges of all decision variables.

Remark 1 Clearly, solving problem $[\mathcal{P}_1]$ directly is very challenging because (i) $[\mathcal{P}_1]$ is an integer programming problem with a non-convex objective function and multi-dimensional decision variables, leading to the potential curse of dimensionality; (ii) since both the importance weights of partial-DTs in the federated DT construction and partial-DT preparations at ESs are uncertain and time-varying, it is necessary to develop an online algorithm with the strong exploration capability; (iii) each sensor can be associated with multiple ESs simultaneously, making the traditional ES-sensor association methods, such as the merge-and-split algorithm and Shapley value-based approaches, inapplicable, as they all assume non-overlapping settings;

and (iv) the configuration cost incurred from the dynamic changes of ES-sensor associations for partial-DT creations increases the interdependence of the decisions across different time frames, meaning that the ES-sensor relationship becomes even more complicated in long runs.

3.4 A Game-Theoretic Approach with Two-Sided Matching and Overlapping Coalitions

In this section, we first transform the original problem $[\mathcal{P}_1]$ into a more tractable two-layer hierarchical game consisting of a two-sided matching game and an overlapping coalition formation game. Then, we analyze these games in detail and propose a DRL-based solution to obtain the corresponding equilibrium in an online manner.

3.4.1 *Problem Transformation to a Hierarchical Game*

We observe from the problem formulation in (3.25a)–(3.25g) that solving $[\mathcal{P}_1]$ is equivalent to simultaneously solve the following two problems, i.e.,

$$[\mathcal{P}_2]: \max_{\{x_{c,b}(t)\}_{\forall t}} \lim_{T\to\infty} \frac{1}{T} \sum\nolimits_{t=0}^{T} U_{CS}(t) \tag{3.26}$$

$$\text{s.t.,} \quad (3.3),\ (3.25b),\ (3.25d).$$

$$[\mathcal{P}_3]: \max_{\{y_{b,n}(t), z_{b,n,w}(t), T_b(t)\}_{\forall t}} \lim_{T\to\infty} \frac{1}{T} \sum\nolimits_{t=0}^{T} \sum\nolimits_{b\in\mathcal{B}} U_b(t) \tag{3.27}$$

$$\text{s.t.,} \quad (3.8),\ (3.14),\ (3.25b),\ (3.25c),\ (3.25e),\ (3.25f),\ (3.25g).$$

Here, $[\mathcal{P}_2]$ and $[\mathcal{P}_3]$ can be seen as optimization problems of the cloud and each ES, respectively, including decision variables only related to themselves. While these two problems share a common term in objective functions and the same constraint (3.25b), so that their decisions are correlated. To be more specific, in problem $[\mathcal{P}_2]$, $U_{CS}(t) = \sum_{b\in\mathcal{B}} \xi \sum_{c\in C} x_{c,b}(t) I_c(t) A_{c,b}(t) - \kappa E^{Total}(t)$ calculates the utility of the cloud in each time frame t. In problem $[\mathcal{P}_3]$, $U_b(t) = \xi \sum_{c\in C} x_{c,b}(t) I_c(t) A_{c,b}(t) - \kappa E_b^{Total}(t) - C^{Conf} \mathcal{F}_b(t)$ can be defined as the utility of ES $b \in \mathcal{B}$ in each time frame t, where we set $E_b^{Total}(t) = E_b^{Dtr}(t) + E_b^{Back}(t) + E_b^{Cre}(t) + E_b^{Mtr}(t)$ for simplicity, and $\mathcal{F}_b(t) = \{y_{b,n}(t)\}_{\forall n} \oplus \{y_{b,n}(t-1)\}_{\forall n}$ indicates whether sensors associated with ES $b \in \mathcal{B}$ change from $t-1$ to t.

Note that the equivalency between solving $[\mathcal{P}_1]$ and simultaneously solving $[\mathcal{P}_2]$ and $[\mathcal{P}_3]$ lies in the fact that (i) decision variables in $[\mathcal{P}_2]$ and $[\mathcal{P}_3]$ together are identical to those in $[\mathcal{P}_1]$; (ii) since $C^{Conf} \mathcal{F}(t)$ is irrelevant to the decision $\{x_{c,b}(t)\}_{\forall t}$ in $[\mathcal{P}_2]$, the optimal solution that maximizes $\lim_{T\to\infty} \frac{1}{T} \sum_{t=0}^{T} U_{CS}(t)$ in $[\mathcal{P}_2]$ also maximizes $\mathcal{U}_{Sys}$ in $[\mathcal{P}_1]$, which can be written as $\mathcal{U}_{Sys} = \lim_{T\to\infty} \frac{1}{T}\sum_{t=0}^{T} U_{CS}(t) - C^{Conf}\mathcal{F}(t)$; and (iii) since $E_{CS}^{I}(t)$ in the objective function of $[\mathcal{P}_1]$ is irrelevant to the decision $\{y_{b,n}(t), z_{b,n,w}(t), T_b(t)\}_{\forall t}$ in $[\mathcal{P}_3]$, the optimal solution that maximizes $\lim_{T\to\infty} \frac{1}{T}\sum_{t=0}^{T}\sum_{b\in\mathcal{B}} U_b(t)$ in $[\mathcal{P}_3]$ also maximizes $\mathcal{U}_{Sys}$ in $[\mathcal{P}_1]$, which can be written as $\mathcal{U}_{Sys} = \lim_{T\to\infty} \frac{1}{T}\sum_{t=0}^{T}\sum_{b\in\mathcal{B}} U_b(t) - \kappa E_{CS}^{I}(t)$.

With the above transformation, we now aim to solve problems $[\mathcal{P}_2]$ and $[\mathcal{P}_3]$. To this end, we consider that the cloud acts as a dummy agent for solving $[\mathcal{P}_2]$ in strategically determining the optimal partial-DT assignments, while each ES acts as a dummy agent for solving $[\mathcal{P}_3]$ in strategically determining the optimal ES-sensor associations along with the optimal computation and communication resource allocations for partial-DT creations. Although these two decision-making processes are tightly coupled, it is clear that when ESs solve the ES-sensor associations and resource allocations, the partial-DT assignments have to be given in advance by the cloud, meaning that ESs can only make decisions after the cloud. This establishes a natural leader-follower relationship between the cloud and ESs in each time frame t. Therefore, as shown in Fig. 3.2, we formulate a two-layer hierarchical game $\mathcal{G}^H$ to solve $[\mathcal{P}_2]$ and $[\mathcal{P}_3]$ over the long run, where (i) in the upper layer, a two-sided matching game $\mathcal{G}_t^{Up}$ is formulated to describe the cloud's problem $[\mathcal{P}_2]$ in each time frame t; and (ii) in the lower layer, an overlapping coalition formation game $\mathcal{G}_t^{Low}$ is formulated to describe the ESs' problems $[\mathcal{P}_3]$ in each time frame t. Unlike traditional hierarchical games in the existing literature [19, 50], where all players' strategies are static, we let $\mathcal{G}^H$ evolve frame-by-frame, enabling the dynamic strategy adjustments. Specifically, in each time frame t, $\mathcal{G}_t^{Up}$ acts as the leader to determine the short-term decision $x_{c,b}(t)$, and based on this, $\mathcal{G}_t^{Low}$ acts as a follower to determine the short-term decisions $y_{b,n}(t)$, $z_{b,n,w}(t)$, and $T_b(t)$. Note that such a process repeats from one frame to another (as will be explained later), integrating these two short-term subgames into a long-term hierarchical game framework of $\mathcal{G}^H$.

Definition 3.1 (*Two-layer hierarchical game $\mathcal{G}^H$ integrated with $\mathcal{G}_t^{Up}$ and $\mathcal{G}_t^{Low}$*) Formally, $\mathcal{G}^H$ can be defined as

$$\mathcal{G}^H = \left\langle G^H, \{\mathcal{G}_t^{Up}, \Pi_t^{Up}\}_{\forall t}, \{\mathcal{G}_t^{Low}, \Pi_t^{Low}\}_{\forall t}, \mathcal{U}_{CS}, \mathcal{U}_{\mathcal{B}} \right\rangle, \tag{3.28}$$

where $G^H = CS \cup \mathcal{B}$ represents the set of participants in $\mathcal{G}^H$, i.e., the cloud CS and all ESs $\mathcal{B}$; $\mathcal{G}_t^{Up}$ stands for the upper-layer subgame in each time frame t, with Π_t^{Up} being the corresponding strategy, and $\mathcal{G}_t^{Low}$ is the lower-layer subgame in each time frame t, with Π_t^{Low} being the corresponding strategy; $\mathcal{U}_{CS} = \{U_{CS}(t)\}_{\forall t}$ and $\mathcal{U}_{\mathcal{B}} = \{U_b(t)\}_{\forall b,t}$ include the utilities of the cloud and ESs in all time frames, respectively.

Definition 3.2 (*Upper-layer subgame: two-sided matching game $\mathcal{G}_t^{Up}$*) For solving $[\mathcal{P}_2]$ in each time frame t, the decision on partial-DT assignments $x_{c,b}(t), \forall c \in C, \forall b \in \mathcal{B}$, can actually be viewed as a bidirectional matching between partial-DTs in C and ESs in $\mathcal{B}$, such that each partial-DT has different preferences for being created on different ESs and vice versa [30]. Hence, $\mathcal{G}_t^{Up}$ can be defined as

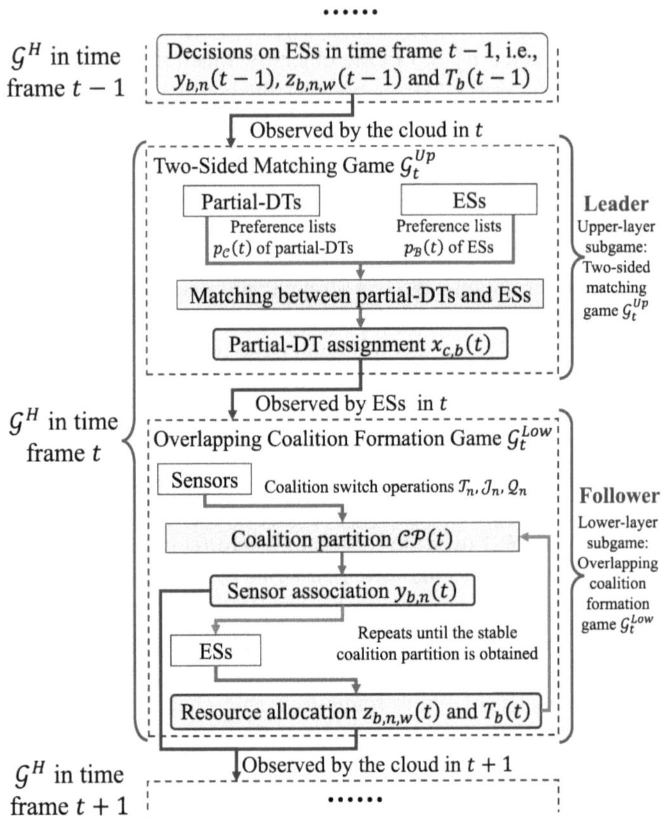

Fig. 3.2 An illustration of the repeated decision-making process of $\mathcal{G}^H$

$$\mathcal{G}_t^{Up} = \left\langle G^{Up}, \Pi_t^{Up} = \{\Phi_t(i)\}_{\forall i \in G^{Up}}, \mathbf{p}, U_{CS}(t), \{U_b(t)\}_{\forall b} \right\rangle, \tag{3.29}$$

where $G^{Up} = \mathcal{C} \cup \mathcal{B}$ denotes all partial-DTs and ESs involved in $\mathcal{G}_t^{Up}$; $\mathbf{p} = \{\mathbf{p}_c(t) = \{\{p^{c,b}\}_{\forall b}\}, \mathbf{p}_b(t) = \{p^{b,c}\}_{\forall b}\}_{\forall c,b}$ are the sets of all partial-DTs' and ESs' preference lists in time frame t, respectively, with $p^{c,b}$, $p^{b,c} \in [0, 1]$ being the ranking value; $\Phi_t(i)$, $\forall i \in G^{Up}$ is the bidirectional matching between $\mathcal{C}$ and $\mathcal{B}$ in each time frame t, i.e.,

$$\Phi_t(c) = b \Leftrightarrow \Phi_t(b) = c, \text{ if and only if } x_{c,b}(t) = 1, \forall c \in \mathcal{C}, \forall b \in \mathcal{B}. \tag{3.30}$$

Definition 3.3 (*Lower-layer subgame: overlapping coalition formation game* $\mathcal{G}_t^{Low}$) For solving $[\mathcal{P}_3]$ in each time frame t, the decisions on ES-sensor associations $y_{b,n}(t)$ and resource allocations $z_{b,n,w}(t)$, $T_b(t)$ can actually be viewed as an overlapping coalition formation among sensors and ESs. Specifically, all sensors are grouped into multiple overlapping coalitions, each of which associated with one ES, and each ES determines $z_{b,n,w}(t)$, $T_b(t)$ based on the state of all coalitions. Sensors may strategically adjust their coalitions based on $z_{b,n,w}(t)$, $T_b(t)$, and this process is conducted in an iterative manner till the convergence. Hence, we define $\mathcal{G}_t^{Low}$ as

$$\mathcal{G}_t^{Low} = \left\langle G^{Low}, \Delta, \Pi_t^{Low} = \left\{\mathcal{CP}(t), \{\pi_b^R(t)\}_{\forall b}\right\}, \{U_b^t(Co_b)\}_{\forall b}, \{U_n(t)\}_{\forall n}\right\rangle, \tag{3.31}$$

where $G^{Low} = \mathcal{N} \cup \mathcal{B}$ denotes all sensors and ESs involved in $\mathcal{G}_t^{Low}$; Δ stands for the set of possible overlapping coalitions; $\mathcal{CP}(t) = \{Co_b(t)\}_{\forall b \in \mathcal{B}}$ represents the coalition partition in time frame t, with $Co_b(t) = \{n \,|\, y_{b,n}(t) = 1, \forall n \in \mathcal{N}\}$ being the coalition for ES $b \in \mathcal{B}$; $\pi_b^R(t) = \{z_{b,n,w}(t), T_b(t)\}_{\forall n,w}$ is the resource allocation strategy of ES $b \in \mathcal{B}$ in time frame t; $U_b^t(Co_b) = U_b(t)$, if $Co_b = Co_b(t)$, shows the utility of ES $b \in \mathcal{B}$ in time frame t given that its associated sensor coalition is Co_b, $U_n(t) = \sum_{b \in \mathcal{B}} y_{b,n}(t) U_n^{Co_b}(t)$ can be seen as the coalitional utility of sensor $n \in \mathcal{N}$ in time frame t, with $U_n^{Co_b}(t)$ being the contribution of sensor n to coalition $Co_b(t)$, calculated as the difference between the utility of ES b with and without sensor n in its associated coalition Co_b, i.e.,

$$\begin{aligned}
U_n^{Co_b}(t) &= U_b^t(Co_b) - U_b^t(Co_b/n) \\
&= x_{c,b}(t)\xi I_c(t)\Big[\Gamma\Big(\sum\nolimits_{i=1}^{t} d_c(i)\Big)\Big(1 - 2^{(T_b(t)(L\delta-2)\delta\gamma)/2}\Big) \\
&\quad - \Gamma\Big(\sum\nolimits_{i=1}^{t} d_c(i) - d_{n,c}(t)\Big)\Big(1 - 2^{\frac{T_b(t)T_{b/n}^*(t)(L\delta-2)\delta\gamma}{2T_b^*(t)}}\Big)\Big] \\
&\quad - \kappa\Big[\rho_b(F_b)^2\Upsilon_b\Big(T_b(t) - T_b(t)T_{b/n}^*(t)/T_b^*(t)\Big) \\
&\quad + P_n x_{c,b}(t) d_{n,c}(t)\Big/R_{n,b}^T(t)\Big] - C^{Conf}\mathcal{F}_b(t)/|\mathcal{B}|.
\end{aligned} \tag{3.32}$$

3.4.2 Game Analysis

Let $(\Pi_t^{Up*}, \Pi_t^{Low*})_{\forall t}$ be the best response of the cloud and ESs in $\mathcal{G}^H$, i.e., the optimal strategies in solving $[\mathcal{P}_2]$ and $[\mathcal{P}_3]$ in every time frame t, then the equilibrium of $\mathcal{G}^H$ can be defined as follows.

Definition 3.4 (*Equilibrium of hierarchical game* $\mathcal{G}^H$) In $\mathcal{G}^H$, a strategy profile $\left(\Pi_t^{Up*}, \Pi_t^{Low*}\right)_{\forall t}$ is the equilibrium if Π_t^{Up*} and Π_t^{Low*} are respectively the equilibria of the upper-layer and lower-layer games in each time frame t, and moreover

$$\begin{aligned}
&\frac{1}{T}\sum\nolimits_{t=1}^{T} U_{CS}(t)|\Pi_t^{Up*}, \Pi_t^{Low*} \geq \frac{1}{T}\sum\nolimits_{t=1}^{T} U_{CS}(t)|\Pi_t^{Up}, \Pi_t^{Low*}, \text{ and} \\
&\frac{1}{T}\sum\nolimits_{t=1}^{T} U_b(t)|\Pi_t^{Low*}, \Pi_t^{Up*} \geq \frac{1}{T}\sum\nolimits_{t=1}^{T} U_b(t)|\Pi_t^{Low}, \Pi_t^{Up*}, \ \forall b \in \mathcal{B}.
\end{aligned} \tag{3.33}$$

Clearly, when such an equilibrium is reached, the long-term utilities of both the cloud and ESs can be maximized so that $[\mathcal{P}_2]$ and $[\mathcal{P}_3]$ are solved, and no one will deviate this equilibrium unilaterally as it also guarantees the optimality in every time frame.

Definition 3.5 (*Equilibrium of the upper-layer subgame $\mathcal{G}_t^{Up}$*) Given the equilibrium Π_t^{Low*} of lower-layer subgame $\mathcal{G}_t^{Low}$, in $\mathcal{G}_t^{Up}$, the equilibrium Π_t^{Up*} is defined as a stable matching $\{\Phi_t^*(i)\}_{\forall i \in G^{Up}}$, which satisfies the condition that no blocking pair (c, b) exists for any partial-DT $c \in C$ and ES $b \in \mathcal{B}$. Particularly, a pair (c, b) is blocking if both conditions $p^{c,b} > p^{c,\Phi_t(c)}$ and $p^{b,c} > p^{b,\Phi_t(b)}$ hold simultaneously, where $\Phi_t(c)$ and $\Phi_t(b)$ represent the current matched partners of c and b in time frame t, respectively.

Definition 3.6 (*Equilibrium of the lower-layer subgame $\mathcal{G}_t^{Low}$*) In $\mathcal{G}_t^{Low}$, the equilibrium Π_t^{Low*} is defined as the best response $\left(\mathcal{CP}^*(t), \{\pi_b^{R*}\}_{\forall b}\right)$, satisfying

$$\begin{aligned} &U_b(t)|\mathcal{CP}^*(t), \{\pi_b^{R*}(t)\}_{\forall b} \geq U_b(t)|\mathcal{CP}(t), \{\pi_b^{R*}(t)\}_{\forall b}, \text{ and} \\ &U_b(t)|\{\pi_b^{R*}(t)\}_{\forall b}, \mathcal{CP}^*(t) \geq U_b(t)|\pi_b^R(t), \{\pi_{b'}^{R*}(t)\}_{\forall b' \neq b}, \mathcal{CP}^*(t), \ \forall b \in \mathcal{B}, \end{aligned} \tag{3.34}$$

where $\mathcal{CP}^*(t) = \{Co_b^*(t)\}_{\forall b \in \mathcal{B}}$ represents the stable overlapping coalition partition of all sensors in time frame t. Here, we say the overlapping coalition partition $\mathcal{CP}^*(t)$ is stable if and only if the total utility of all ESs cannot be improved by unilaterally altering the coalition of any sensor even in an overlapped manner.

Definitions 3.4, 3.5 and 3.6 reveal an interdependence among $\mathcal{G}^H$, $\mathcal{G}_t^{Up}$, and $\mathcal{G}_t^{Low}$, such that the equilibrium of $\mathcal{G}^H$ depends on the equilibria of both $\mathcal{G}_t^{Up}$ and $\mathcal{G}_t^{Low}$ in each time frame t, while also extends them to accommodate to dynamic settings. Therefore, to facilitate the analysis, we first prove the existence of the equilibria in $\mathcal{G}_t^{Up}$ and $\mathcal{G}_t^{Low}$, and in turn prove the existence of the equilibrium in $\mathcal{G}^H$.

Lemma 3.1 (Existence of the equilibrium in upper-layer subgame $\mathcal{G}_t^{Up}$) *Suppose the equilibrium Π_t^{Low*} of lower-layer subgame $\mathcal{G}_t^{Low}$ exists, $\mathcal{G}_t^{Up}$ has at least one equilibrium Π_t^{Up*}, i.e., the stable matching $\{\Phi_t(i)\}_{\forall i \in G^{Up}}$ in any time frame t.*

Proof Since running the lower-layer subgame $\mathcal{G}_t^{Low}$ requires the partial-DT assignments from the upper-layer game $\mathcal{G}_t^{Up}$, a leader-follower relationship exists between these two games. In other words, the existence of the equilibrium in $\mathcal{G}_t^{Up}$ depends on that in $\mathcal{G}_t^{Low}$ [19, 50].

Suppose the equilibrium of $\mathcal{G}_t^{Low}$ exists, $\mathcal{G}_t^{Up}$ is an one-to-one matching game with (i) a finite set of participants C and $\mathcal{B}$, where the total amount is $|C| + \mathcal{B}$; and (ii) a finite set of strategy $\Pi_t^{Up} = \{\Phi_t(i)\}_{\forall i \in G^{Up}}$, where the maximum cardinality is $\binom{|\mathcal{B}|}{|C|} \cdot |C|!$. Therefore, $\mathcal{G}_t^{Up}$ is a finite game which definitely has a Nash equilibrium according to [21].

Lemma 3.2 (Existence of the equilibrium in lower-layer subgame $\mathcal{G}_t^{Low}$) *$\mathcal{G}_t^{Low}$ has at least one equilibrium Π_t^{Low*}, i.e., a stable coalition partition $\mathcal{CP}^*(t)$ followed by optimal resource allocations $\{\pi_b^{R*}\}_{\forall b}$ in any time frame t.*

Proof First, we prove that, given the optimal resource allocations $\{\pi_b^{R*}\}_{\forall b}$, the stable coalition partition $\mathcal{CP}^*(t)$ exists. Considering that coalitions are formed in an overlapped manner, there are three possible cases for each sensor $n \in \mathcal{N}$ to be determined in changing its coalition, i.e., (i) n leaves one of its current coalitions and joins another one; (ii) n directly joins another coalition as an overlap; and (iii) n chooses to leave one of its current coalitions unilaterally. In each case, we prove that $\mathcal{G}_t^{Low}$ is an exact potential game (EPG) [9], which ensures the existence of at least one pure Nash equilibrium or equivalently a stable overlapping coalition partition. To this end, we can design a potential function as

$$\zeta^t(Co_a, Co_{-a}) = \sum\nolimits_{b\in\mathcal{B}} U_b(t),$$

where Co_{-a} is set of all sensor coalitions of ESs other than $a \in \mathcal{B}$. Hereafter, we analyze all three cases for each sensor n and show that the change in the coalitional utility function $U_n(t)$ equals the change in potential function ζ^t.

Case 1: Suppose that sensor $n \in \mathcal{N}$ is determined to leave coalition Co_a and join another coalition $Co_{a'}$, $\forall a, a' \in \mathcal{B}$, the difference brought to the utility of n, i.e., $U_n(t)$, can be calculated as

$$\begin{aligned}
&U_n(t)\big|Co_a, Co_{a'} - U_n(t)\big|Co_a\backslash n, Co_{a'} \cup n\\
&= U_n^{Co_a}(t) + U_n^{others}(t) - \big(U_n^{Co_{a'}\cup n}(t) + U_n^{others}(t)\big)\\
&= U_a^t(Co_a) - U_a^t(Co_a\backslash n) - \big(U_{a'}^t(Co_{a'} \cup n) - U_{a'}^t(Co_{a'})\big),\\
&= \sum\nolimits_{b\in\mathcal{B}} U_b(t)\big|Co_a, Co_{a'} - \sum\nolimits_{b\in\mathcal{B}} U_b(t)\big|Co_a\backslash n, Co_{a'} \cup n,\\
&= \zeta^t(Co_a, Co_{a'}, Co_{-a,a'}) - \zeta^t(Co_a\backslash n, Co_{a'} \cup n, Co_{-a,a'}),
\end{aligned}$$

where $U_n^{others}(t)$ represents the total utility that n obtains in other coalitions, $U_{-a,a'}^t$ and $Co_{-a,a'}$ denote the total utility and all coalitions of ESs other than a and a', respectively. Therefore, in this case, the difference in $U_n(t)$ equals to that in ζ^t.

Case 2: Suppose that sensor $n \in \mathcal{N}$ is determined to join coalition Co_a, $\forall a \in \mathcal{B}$ as an overlap, the difference brought to $U_n(t)$ can be calculated as

$$\begin{aligned}
&U_n(t) - U_n(t)\big|Co_a \cup n\\
=&U_n^{others}(t) - \big(U_n^{Co_a\cup n}(t) + U_n^{others}(t)\big)\\
=&U_a^t(Co_a) - U_a^t(Co_a \cup n),\\
=&U_a^t(Co_a) + U_{-a}^t - \big(U_a^t(Co_a \cup n) + U_{-a}^t\big),\\
=&\sum\nolimits_{b\in\mathcal{B}} U_b(t)\big|Co_a - \sum\nolimits_{b\in\mathcal{B}} U_b(t)\big|Co_a \cup n,\\
=&\zeta^t(Co_a, Co_{-a}) - \zeta^t(Co_a \cup n, Co_{-a}).
\end{aligned}$$

Therefore, in this case, the difference in $U_n(t)$ equals to that in ζ^t.

Case 3: Suppose that sensor $n \in \mathcal{N}$ is determined to unilaterally quit its current coalition Co_a, $\forall a \in \mathcal{B}$. Obviously, this is equivalent to that sensor n is determined to leave Co_a and join an empty coalition $\varnothing$, and the difference in $U_n(t)$ equals to that in ζ^t following the proof in case 1.

Therefore, following [9], the lower-layer subgame $\mathcal{G}_t^{Low}$ is an EPG. Since it is widely proved that EPG has at least one pure Nash equilibrium, we can conclude that the equilibrium of $\mathcal{G}_t^{Low}$ (i.e., the stable overlapping coalition partition $\mathcal{CP}^*(t)$) in each time frame exists.

Finally, we prove that the best response $\left(\mathcal{CP}^*(t), \{\pi_b^{R*}\}_{\forall b}\right)$ exists. Given a stable coalition partition $\mathcal{CP}^*(t)$, the number of possible resource allocation strategies π_b^{R*} for each ES $b \in \mathcal{B}$ is $|\mathcal{N}||\mathcal{W}| \left\lceil \frac{2}{(L\delta-2)\delta\gamma} \log_2(1 - \frac{A_c^{Req}}{\Gamma\left(\sum_{i=1}^{t} \min_{n,c} d_{n,c}(t)\right)}) \right\rceil$, implying that $\mathcal{G}_t^{Low}$ is also a finite game. According to [21], finite games always have a Nash equilibrium. Therefore, $\mathcal{G}_t^{Low}$ has a Nash equilibrium, represented by the pair $\left(\mathcal{CP}^*(t), \{\pi_b^{R*}\}_{\forall b}\right)$. This completes the proof. □

Theorem 3.1 (Existence of the equilibrium in hierarchical game $\mathcal{G}^H$) *In $\mathcal{G}^H$, there always exists an equilibrium $(\Pi_t^{Up*}, \Pi_t^{Low*})_{\forall t}$ in long runs.*

Proof Based on Lemmas 3.1 1 and 3.2, we aim to prove that $\mathcal{G}^H$ is a multi-player, general-sum, discounted-reward stochastic game, which naturally has a Markov perfect equilibrium (MPE) [3, 37].

First, we show that the strategy space for both the cloud and each ES is finite across multiple time frames in $\mathcal{G}^H$. In terms of the cloud, the maximum cardinality of $\{\Phi_t^*(i)\}_{\forall i \in G^{Up}}$ is $|C|! \cdot \binom{|\mathcal{B}|}{|C|}$. Meanwhile, in terms of each ES, the maximum cardinality of $z_{b,n,w}(t)$ is $|\mathcal{N}||\mathcal{W}|$, the maximum cardinality of $T_b(t)$ can be calculated as $\left\lceil \frac{2}{(L\delta-2)\delta\gamma} \log_2(1 - \frac{A_c^{Req}}{\Gamma\left(\sum_{i=1}^{t} \min_{n,c} d_{n,c}(t)\right)}) \right\rceil$, and the maximum cardinality of $\mathcal{CP}^*(t)$ can be denoted as $2^{|\mathcal{B}||\mathcal{N}|}$. These indicate that the strategy spaces in $\mathcal{G}^H$ are all finite across multiple time frames.

Furthermore, since strategies of the cloud and ESs in each time frame depend on themselves only and the resulted system states, e.g., the coalition structure and model accuracy, the state transitions in $\mathcal{G}^H$ satisfy the Markov property. Besides, it is clear that $\mathcal{G}^H$ is not a zero-sum game, as the total utility of the cloud and ESs does not sum to zero. Then, according to [3, 37], we can conclude that $\mathcal{G}^H$ with finite strategy spaces is a multi-player, general-sum stochastic game.

In addition, referring to [16, 18], maximizing the average infinite time-horizon utility and discounted infinite time-horizon utility are equivalent, and thus we have

$$\frac{1}{T}\sum\nolimits_{t=1}^{T} U_x(t) \Leftrightarrow \sum\nolimits_{t=0}^{T} \eta^t U_x(t), \forall x \in \{CS \cup \mathcal{B}\},$$

where η stands for the discount factor to balance the instant and future utilities. Given this fact, we can further conclude that $\mathcal{G}^H$ is a multi-player, general-sum, discounted-reward stochastic game in long runs, which must have an MPE. Furthermore, taking into account (3.34) and the equilibrium in the upper-layer subgame, such MPE can be written as

$$\sum_{t=1}^{T} \eta^t U_{CS}(t)|\Pi_t^{Up*}, \Pi_t^{Low*} \geq \sum_{t=1}^{T} \eta^t U_{CS}(t)|\Pi_t^{Up}, \Pi_t^{Low*}, \text{ and}$$
$$\sum_{t=1}^{T} \eta^t U_b(t)|\Pi_t^{Low*}, \Pi_t^{Up*} \geq \sum_{t=1}^{T} \eta^t U_b(t)|\Pi_t^{Low}, \Pi_t^{Up*}, \; \forall b \in \mathcal{B}, \tag{3.35}$$

meaning that (3.33) is satisfied. This completes the proof. □

3.4.3 *A DRL-Based Solution Integrated with Stable Matching and Overlapping Coalition Formation (DMO)*

In this subsection, we develop an SOCF algorithm and leverage the GS algorithm to find the equilibria of lower-layer subgame $\mathcal{G}_t^{Low}$ and upper-layer subgame $\mathcal{G}_t^{Up}$ within each single time frame t, respectively. We then propose a DRL-based solution to extend these two short-term equilibria into the long-term equilibrium of hierarchical game $\mathcal{G}^H$. We name such a solution approach as DMO, with its overall structure illustrated by Fig. 3.3. Specifically, the proposed DMO operates in two stages. In Stage 1, given any partial-DT assignments $x_{c,b}(t)$, the initial resource allocations $z_{b,n,w}(t)$ and $T_b(t)$ are derived by the actor networks of proximal policy optimization (PPO)-based agents in the DRL framework and then inputted into SOCF algorithm to obtain a temporary stable coalition partition $\mathcal{CP}^*(t)$. Iteratively, $\mathcal{CP}^*(t)$ is fed back into the PPO-based agents to update $\{\pi_b^*(t)\}_{\forall b}$ until the equilibrium of $\mathcal{G}_t^{Low}$, i.e., $(\mathcal{CP}^*(t), \{\pi_b^*(t)\}_{\forall b})$, converges. In Stage 2, GS algorithm is employed to acquire the stable matching $\{\Phi_t^*(i)\}_{\forall i \in G^{Up}}$, or equivalently the optimal partial-DT assignments $x_{c,b}^*(t)$, i.e., the equilibrium of $\mathcal{G}_t^{Up}$. This is achieved by inputting the preference lists $\{\mathbf{p}_c(t)\}_{\forall c}$ and $\{\mathbf{p}_b(t)\}_{\forall b}$, which are dynamically generated by the PPO-based agents in the DRL framework, informed by their observations of the system states. The whole process repeats frame-by-frame, allowing PPO-based agents to dynamically exploit historical knowledge and optimize strategies of both the cloud and ESs to achieve global optima in long runs, eventually reaching the equilibrium of $\mathcal{G}^H$ and thus solving the original problem.

3.4.3.1 Short-Term Equilibrium Solutions to Subgames

In each time frame t, the solution to $\mathcal{G}_t^{Low}$'s equilibrium $\left(\mathcal{CP}^*(t), \{\pi_b^{R*}\}_{\forall b}\right)$ should be first obtained due to the hierarchy that $\mathcal{G}_t^{Low}$ is the follower of $\mathcal{G}_t^{Up}$. However, finding $\mathcal{CP}^*(t)$ is much more difficult than traditional coalition games [9, 38], because in the considered system, sensors are allowed to form multiple overlapping coalitions, of which conditions for coalition partitions are hard to be evaluated. To this end, we extend the conventional coalition switch rules [38] into overlapping coalition switch rules by particularly considering the condition that a sensor can join multiple coalitions as an overlap, and reformulate the utilities of overlapping coalition members. As such, we develop an SOCF algorithm to optimize $\mathcal{CP}^*(t)$ in one time frame t,

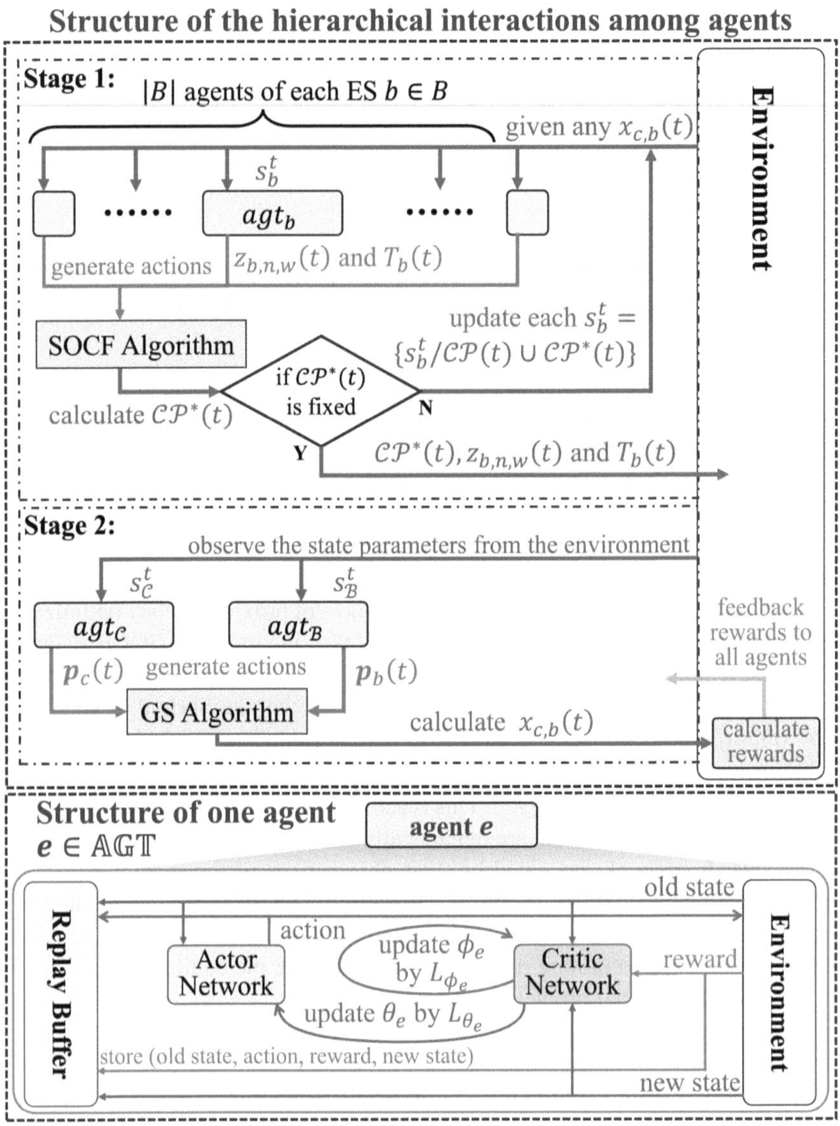

Fig. 3.3 Overview of the proposed DMO approach

producing the optimal $y_{b,n}(t)$, given the optimal $\{\pi_b^{R*}(t)\}_{\forall b}$ (derived by the actor network of PPO-based agents dynamically, which will be presented later). Note that the finally converged $\mathcal{CP}^*(t)$ and $\{\pi_b^{R*}(t)\}_{\forall b}$ together constitute $\mathcal{G}_t^{Low}$'s equilibrium introduced in Definition 3.6.

In $\mathcal{G}_t^{Low}$, the fundamental operations of a sensor in changing its coalition in an overlapped manner can be categorized into three switch operations, i.e., transferring operation, joining operation and quitting operation. Accordingly, we define transferring rule, joining rule and quitting rule to provide their condition of the sensors' choosing to perform these switch operations as follows. From ES's perspective, associating a sensor may decrease its utility when the costs stem form data preprocessing and transmitting overwhelm the benefit from enhanced model quality of partial-DT. In this case, from sensor's perspective, it may obtain negative utility from joining a coalition. Therefore, the sensor has strong motivation to consider transferring from one coalition to another or joining a new one to increase its utility or only quit the current coalition if there exits no coalition to increase its utility. Following the game formulations in [19, 45], these fundamental operations of a sensor in $\mathcal{G}_t^{Low}$ can be categorized into transferring operation, joining operation, and quitting operation, and each kind of operations will lead to a change in overlapping coalition partition. Accordingly, we define transferring rule, joining rule and quitting rule to provide the condition of the sensors' choosing to leave or join a coalition in an overlapped manner.

Definition 3.7 (*Overlapping Coalition switch rules*) Let $Co_a, Co_b \in \Delta$ respectively denote the coalitions for ESs $a \in \mathcal{B}$ and $b \in \mathcal{B}, a \neq b$, in $\mathcal{G}_t^{Low}$, a sensor $n \in Co_a, n \notin Co_b$ can be determined to change its coalition in an overlapped manner by (i) transferring from Co_a to Co_b, denoted as $\mathcal{T}_n(Co_a, Co_b)$, (ii) joining Co_b as an overlap, denoted as $\mathcal{J}_n(Co_b)$, and (iii) unilaterally quitting Co_a, denoted as $\mathcal{Q}_n(Co_a)$, of which the conditions are as follows:

(1) *Transferring rule*: $\mathcal{T}_n(Co_a, Co_b)$ is performed only when (i) the remaining subcarriers of ES b is sufficient, i.e., (3.14) holds; (ii) the utility of n can be increased, i.e.,

$$U_n^{Co_b \cup n}(t) \geq \max\{0, U_n^{Co_a}(t)\}; \tag{3.36}$$

(iii) the utility of any sensors except n in Co_b cannot be decreased, i.e.,

$$U_{\bar{n}}^{Co_b \cup n}(t) \geq U_{\bar{n}}^{Co_b}(t), \forall \bar{n} \in Co_b; \tag{3.37}$$

and (iv) the utility of any sensors in other coalitions that n joined as an overlap cannot be decreased, i.e.,

$$\begin{aligned} &U_{\bar{n}}^{Co_e}(t) \,\Big|\, \mathcal{CP}(t) = \{Co_a/i, Co_b \cup n, Co_{-a,b}\} \geq \\ &U_{\bar{n}}^{Co_e}(t) \,\Big|\, \mathcal{CP}(t) = \{Co_a, Co_b, Co_{-a,b}\}, \\ &\forall \bar{n} \in Co_e, \forall Co_e \in Co_{-a,b}, x_{e,n} = 1, \end{aligned} \tag{3.38}$$

where $Co_{-a,b}$ stands for all coalitions in $\mathcal{CP}(t)$ other than Co_a and Co_b.

(2) *Joining rule*: $\mathcal{J}_n(Co_b)$ is performed only when (i) constraints (3.8), (3.14) and (3.37) hold; (ii) the utility of n can be increased, i.e.,

$$U_n^{Co_b \cup n}(t) \geq 0; \tag{3.39}$$

and (iii) the utility of any sensors in other coalitions that n joined as an overlap cannot be decreased, i.e.,

$$U_{\bar{n}}^{Co_e}(t)\Big|\mathcal{CP}(t) = \{Co_b \cup n, Co_{-b}\} \geq U_{\bar{n}}^{Co_e}(t)\Big|\mathcal{CP}(t) = \{Co_b, Co_{-b}\},$$
$$\forall \bar{n} \in Co_e, \forall Co_e \in Co_{-b}, x_{e,n} = 1. \tag{3.40}$$

(3) *Quitting rule*: $\mathcal{Q}_n(Co_a)$ is performed only when (i) the utility of n can be increased, i.e.,

$$U_n^{Co_a}(t) \leq 0; \tag{3.41}$$

(ii) the utility of other sensors in Co_a cannot be decreased, i.e.,

$$U_{\bar{n}}^{Co_a/n}(t) \geq U_{\bar{n}}^{Co_a}(t), \forall \bar{n} \in Co_a. \tag{3.42}$$

We are now able to design an SOCF algorithm based on the above coalition switch rules to obtain the stable coalition partition $\mathcal{CP}^*(t)$ in lower-layer subgame $\mathcal{G}_t^{Low}$. Specifically, in each time frame t, given the optimal resource allocations $\{\pi_b^{R*}(t)\}_{\forall b}$ and partial-DT assignments $x_{c,b}(t)$, each ES initializes its ES-sensor association by (i) calculating the set of eligible sensors $\mathbb{N}_b(t) = \{n \mid R_{n,b}^T(t) \geq R_{Min}^T\}$, where each sensor satisfies (3.25c); (ii) selecting all sensors in its previous coalition $Co_b(t-1)$; and (iii) randomly selecting at most $\max\{|\mathcal{W}| - |Co_b(t-1)|\}$ sensors in $\mathbb{N}_b(t)$ without violating (3.8) and (3.14). Then the initial $\mathcal{CP}(t)$ is formed, and $|\mathcal{N}|$ sensors are required to perform coalition switch operations. After initialization, each sensor simultaneously performs switch operations, i.e., $\mathcal{T}_n(Co_a, Co_b), \mathcal{J}_n(Co_b)$ and $\mathcal{Q}_n(Co_a)$, adhering to the switch rules defined in Definition 3.7. This process repeats until no sensor has the motivation to perform any further switch operations, i.e., the coalition partition converges to $\mathcal{CP}^*(t)$. Correspondingly, the optimal ES-sensor associations, i.e., $y_{b,n}^*(t), \forall b \in \mathcal{B}, \forall n \in \mathcal{N}$, can be obtained from $\mathcal{CP}^*(t)$. The detailed procedure of this algorithm is summarized in Algorithm 1.

Theorem 3.2 (Convergence and optimality of SOCF) *Starting from any initial overlapping coalition partition, the proposed SOCF algorithm can eventually converge to a stable overlapping coalition partition $\mathcal{CP}^*(t)$ in each time frame t, yielding the optimal ES-sensor associations.*

Proof In the proposed SOCF algorithm, each sensor $n \in \mathcal{N}$ is asked to perform the switch operations, i.e., $\mathcal{T}_n(Co_a, Co_b), \mathcal{J}_n(Co_b)$ or $\mathcal{Q}_n(Co_a)$, following the switch

Algorithm 1: SOCF Algorithm for $\mathcal{CP}^*(t)$

Input: Partial-DT assignment $x_{c,b}(t)$, initial decisions on $z_{b,n,w}(t)$ and $T_b(t)$, previous decision on $y_{b,n}(t-1)$;
Output: Stable coalition partition $\mathcal{CP}^*(t)$, or equivalently the optimal ES-sensor associations $y^*_{b,n}(t)$;

for *ES* $b \in \mathcal{B}$ **do**
  ES b calculates $\mathbb{N}_b(t)$ according to constraint (3.25);
  ES b selects all sensors in $Co_b(t-1)$ and randomly selects at most $\max\{|\mathcal{W}| - |Co_b(t-1)|\}$ sensors in $\mathbb{N}_b(t)$.
All ESs generate the initial overlapping coalition partition $\mathcal{CP}(t)$;
while $\mathcal{CP}(t)$ *is different from it in the previous loop* **do**
  for $n \in \mathcal{N}$ **do**
    for $Co_a \in \{Co_x | n \in Co_x\}$ **do**
      for $Co_b \notin \{Co_x | n \in Co_x\}$ **do**
        if *Transferring rule is satisfied* **then**
          $\mathcal{T}_n(Co_a, Co_b)$ is performed;
        else if *Joining rule is satisfied* **then**
          $\mathcal{J}_n(Co_a, Co_b)$ is performed;
        else if *Quitting rule is satisfied* **then**
          $\mathcal{Q}_n(Co_a)$ is performed;
    Update $\mathcal{CP}(t)$, all ESs synchronize such information;
According to final $\mathcal{CP}(t)$, all ESs update $y_{b,n}(t)$.

rules defined in Definition 7. Denote that the coalition partition becomes $\mathcal{CP}^{(k)}(t)$ after k iterations, we prove that the sum utility of all ESs, i.e., $\sum_{\forall b \in \mathcal{B}}$, will not decrease from $\mathcal{CP}^{(k-1)}(t)$ to $\mathcal{CP}^{(k)}(t)$, regardless of which switch operation that sensor n is determined to perform.

Case 1: Suppose sensor $n \in \mathcal{N}$ is determined to perform the transferring operation $\mathcal{T}_n(Co_a, Co_b)$, $\forall a, b \in \mathcal{B}$ in kth iteration. Then, according to the transferring rule, we have

$$
\begin{aligned}
& U_n^{Co_b \cup n}(t) \geq \max\{0, U_n^{Co_a}(t)\} \\
\Leftrightarrow\ & U_b^t(Co_b \cup n) - U_b^t(Co_b) \geq U_b^t(Co_a) - U_b^t(Co_a \backslash n) \\
\Leftrightarrow\ & U_b^t(Co_b \cup n) + U_b^t(Co_a \backslash n) \geq U_b^t(Co_a) + U_b^t(Co_b).
\end{aligned}
$$

This implies that the total utility of coalitions Co_a and Co_b increases if $\mathcal{T}_n(Co_a, Co_b)$ is conducted.

Case 2: Suppose that sensor $n \in \mathcal{N}$ is determined to perform the joining operation $\mathcal{J}_n(Co_b)$, $\forall b \in \mathcal{B}$ in kth iteration. Then, according to the joining rule, we have

$$
U_n^{Co_b \cup n}(t) \geq 0 \Leftrightarrow U_b^t(Co_b \cup n) \geq U_b^t(Co_b).
$$

This indicates that the utility of coalition Co_b increases if $\mathcal{J}_n(Co_b)$ is conducted.

Case 3: Suppose that sensor $n \in \mathcal{N}$ is determined to perform $\mathcal{Q}_n(Co_a), \forall a \in \mathcal{B}$ in kth iteration. Then, according to the quitting rule, we have

$$U_n^{Co_a}(t) \le 0 \Leftrightarrow U_a^t(Co_a) \ge U_a^t(Co_a \backslash n).$$

This shows that the utility of coalition Co_a does not decrease if $\mathcal{Q}_n(Co_a)$ is conducted.

All these three cases demonstrate that whenever a sensor $n \in \mathcal{N}$ is determined to perform a coalition switch operation, the total utility of the coalitions either increases or maintains the same. Additionally, the utilities of all other coalitions will not decrease following Definition 3.7. Therefore, we can conclude that the total utility of all ESs, i.e., $\sum_{\forall b \in \mathcal{B}} U_b(t)$, monotonically increases in any iteration k of SOCF algorithm. Since the total utility of all ESs is bounded by the limited feature data that can be collected in each time frame, the coalition partition $\mathcal{CP}^{(k)}(t)$ is guaranteed to converge after a finite number of iterations.

Then, given the equilibrium solution of $\mathcal{G}_t^{Low}$, for the upper-layer subgame $\mathcal{G}_t^{Up}$, since it is a standard one-to-one matching game, we employ the well-known GS algorithm [29, 30] to achieve the equilibrium, i.e., the stable matching $\{\Phi_t^*(i)\}_{\forall i \in G^{Up}}$ in each time frame t, of which the input is preference lists $\{\mathbf{p}_c(t)\}_{\forall c}$ and $\{\mathbf{p}_b(t)\}_{\forall b}$. As GS algorithm has been widely proved with the guarantee of convergence to a stable matching, $\{\Phi_t^*(i)\}_{\forall i \in G^{Up}}$ can certainly be obtained, leading to the equilibrium of $\mathcal{G}_t^{Up}$.

3.4.3.2 Long-Term Equilibrium Solution to Hierarchical Game

Having the short-term equilibria of subgames $\mathcal{G}_t^{Up}$ and $\mathcal{G}_t^{Low}$ in each time frame t, we can extend them into the long-term equilibrium of hierarchical game $\mathcal{G}^H$. Note that the long-term decision-makings between any two consecutive time frames are inherently coupled, particularly, (i) the decisions on preference orders $\mathbf{p}_c(t), \mathbf{p}_b(t)$ affect the partial-DT assignments $x_{c,b}(t)$ determined by the cloud, which subsequently affects the coalition partition $\mathcal{CP}(t)$ and resource allocation $\pi_b^R(t)$ determined by each ES $b \in \mathcal{B}$; and (ii) the decisions on resource allocations $\{\pi_b^R(t)\}_{\forall b}$ impacts $\mathcal{CP}(t)$ in each time frame t, which in turn affects $\mathbf{p}_c(t+1), \mathbf{p}_b(t+1)$ in the next time frame $t+1$. To tackle such an issue, we first model the long-term decision-makings on resource allocations and preference lists as two coupled Markov decision processes (MDPs) [33, 52], and then develop a DMO approach to dynamically solve these MDPs. Particularly, given the outputs of PPO-based agents, we call SOCF and GS algorithms to derive $\mathcal{CP}^*(t)$ and $\{\Phi_t^*(i)\}_{\forall i \in G^{Up}}$ in each time frame t, and then we continuously train these agents using the historical knowledge, dynamically updating their outputs. The final outputs of DMO, including $\{\pi_b^R(t)\}_{\forall b}$ and $\mathbf{p}_c(t), \mathbf{p}_b(t)$ from PPO-based agents, and $\mathcal{CP}(t)$ and $\{\Phi_t^*(i)\}_{\forall i \in G^{Up}}$ from the SOCF and GS algorithms within the DRL framework, together constitute the long-term equilibrium of $\mathcal{G}^H$, or equivalently the optimal solution to the original problem $[\mathcal{P}_1]$.

In the following, we formulate the MDPs for the decision-makings on resource allocations $\{\pi_b^R(t)\}_{\forall b}$ and preference lists $\mathbf{p}_c(t)$, $\mathbf{p}_b(t)$ as $\mathcal{M}_R = \{\mathcal{S}_R, \mathcal{A}_R, \Xi_R, \mathcal{R}_R\}$ and $\mathcal{M}_{PL} = \{\mathcal{S}_{PL}, \mathcal{A}_{PL}, \Xi_{PL}, \mathcal{R}_{PL}\}$, respectively, in detail.

(1) *Environment States*: In each time frame t, the environment states in MDP $\mathcal{M}_R$ is the observation on system dynamics in the view of each ES $b \in \mathcal{B}$ in $\mathcal{G}_t^{Low}$, i.e., $s_b^t = \left\{\{I_c(t), d_c(t-1), \sum_{i=1}^{t-1} d_c(i)\}_{\forall c}, \{\Phi_t^*(i)\}_{\forall i \in G^{Up}}, \mathcal{CP}(t)\right\}$. Meanwhile, the environment states in MDP $\mathcal{M}_{PL}$ is the observations on system dynamics in the view of all partial-DTs C and all ESs $\mathcal{B}$ in $\mathcal{G}_t^{Up}$, which can be defined as $s_C^t = \left\{\{I_c(t), d_c(t-1), \sum_{i=1}^{t-1} d_c(i)\}_{\forall c}, \{\pi_b^R(t-1)\}_{\forall b}, \mathcal{CP}(t-1)\right\}$, and $s_{\mathcal{B}}^t = \left\{\{I_c(t), d_c(t-1), \sum_{i=1}^{t-1} d_c(i)\}_{\forall c}, \{\Phi_t^*(i)\}_{\forall i \in G^{Up}}\right\}$, respectively. For conciseness, let $\mathcal{S}_{PL} = \{s_C^t, s_{\mathcal{B}}^t\}_{\forall t}$ and $\mathcal{S}_R = \{s_b^t\}_{\forall b,t}$ denote the state space of MDPs $\mathcal{M}_R$ and $\mathcal{M}_{PL}$, respectively.

(2) *Actions*: In each time frame t, the actions in MDPs $\mathcal{M}_R$ and $\mathcal{M}_{PL}$ are exactly the resource allocations and preference lists, and thus we have their respective action spaces $\mathcal{A}_R = \{a_b^t\}_{\forall b,t} = \{z_{b,n,w}(t), T_b(t)\}_{\forall b,t}$ and $\mathcal{A}_{PL} = \{a_C^t, a_{\mathcal{B}}^t\}_{\forall t} = \{\{\mathbf{p}_c(t)\}_{\forall c}, \{\mathbf{p}_b(t)\}_{\forall b}\}_{\forall t}$.

(3) *State Transition Probabilities*: The state transition probability from state $s \in \mathcal{S}_R$ or $\mathcal{S}_{PL}$ to another state $s' \in \mathcal{S}_R$ or $\mathcal{S}_{PL}$ by taking action $a \in \mathcal{A}_R$ or $\mathcal{A}_{PL}$ is expressed as $\Xi^{s,s'} = Pr(s'|s,a) \in [0,1]$.

(4) *Rewards*: The immediate rewards for ES $b \in \mathcal{B}$ in $\mathcal{M}_R$ can be calculated as

$$r_b^t = \begin{cases} U_b(t), \text{if } \tau^{Total}(t) \le \delta_T, \\ 0, \text{otherwise}, \end{cases} \tag{3.43}$$

and that for i, $\forall i = C$ or $\mathcal{B}$ in $\mathcal{M}_{PL}$ can be written as

$$r_i^t = \begin{cases} U_{CS}(t), \text{if } i = C \text{ and if } \tau^{Total}(t) \le \delta_T, \\ \sum_{b \in \mathcal{B}} U_b(t), \text{if } i = \mathcal{B} \text{ and if } \tau^{Total}(t) \le \delta_T, \\ 0, \text{otherwise}. \end{cases} \tag{3.44}$$

For conciseness, let $\mathcal{R}_R = \{r_b^t\}_{\forall b,t}$ and $\mathcal{R}_{PL} = \{r_C^t, r_{\mathcal{B}}^t\}_{\forall t}$ denote the reward spaces in $\mathcal{M}_R$ and $\mathcal{M}_{PL}$, respectively.

Based on the above formulation, we next elaborate the DMO approach in terms of its agent structure, training workflow, and network parameter updates.

First, the decisions in either $\mathcal{M}_R$ or $\mathcal{M}_{PL}$ are multi-dimensional, making it computationally intensive to explore the vast action space. To address this complexity, we assign the actions in $\mathcal{A}_R$ and $\mathcal{A}_{PL}$ to multiple agents, which enhances exploration efficiency and accelerates the training process. Particularly, we introduce $|\mathcal{B}|$ agents agt_b, $\forall b \in \mathcal{B}$, where each agent agt_b generates resource allocation $\pi_b^R(t)$. Additionally, agents agt_C and $agt_{\mathcal{B}}$ are responsible for generating $\{\mathbf{p}_c(t)\}_{\forall c}$ and $\{\mathbf{p}_b(t)\}_{\forall b}$, respectively. The set of all agents is denoted as $\mathbb{AGT}$. To improve the training efficiency, we employ the actor-critic (AC) framework [12]. For each agent $e \in \mathbb{AGT}$, we design (i) a critic network with parameter ϕ_e to estimate state-value $V_{\phi_e}(s) \approx V_e(s)$ of e, where $V_e(s) = \mathbb{E}\{\sum_{t=1}^{T} \eta^t r_e^t | s_0 = s\}$ indicates the long-term discounted reward for e with discount factor η and initial state s_0; (ii) an actor network with parameter θ_e to parameterize the optimal strategy of the corresponding agent; and (iii) a replay buffer to store the historical knowledge, including the previous states, new states, actions and rewards, during the DRL training process.

Second, since the formulated MDPs are inherently coupled, we further propose a hierarchical training process in DRL, where the interaction among all agents adhere to the decision-making sequence in $\mathcal{G}^H$. Specifically, in each time frame t within one training step, such interactions present two stages, (i) in stage 1, the agents $agt_b, \forall b \in \mathcal{B}$ simultaneously generate $\pi_b^R(t)$ based on s_b^t with initial coalition partition $\mathcal{CP}(t-1)$, then we call the SOCF algorithm to iteratively obtain the stable coalition partition and update s_b^t for each agent agt_b, until both $\mathcal{CP}(t)$ and $\pi_b^R(t)$ converge; and (ii) in stage 2, $agt_{\mathcal{C}}$ and $agt_{\mathcal{B}}$ first generate $\{\mathbf{p}_c(t)\}_{\forall c}$ and $\{\mathbf{p}_b(t)\}_{\forall b}$ based on $s_{\mathcal{C}}^t$ and $s_{\mathcal{B}}^t$, respectively, and then we call GS algorithm to obtain $\Phi_t^*(c)$. Note that original reward functions in (3.43), (3.44) may introduce the problem of sparse rewards [22], leading to a lack of sufficient exploration on both the action and state spaces. To handle this, we introduce a penalty for violating constraint (3.25b) to the reward functions, i.e., $\max\left\{\psi\left(\tau^{Total}(t)-\delta_T\right), 0\right\}$, where ψ denotes the penalty coefficient. Accordingly, we can calculate the rewards of all agents in time frame t, and then store the tuple $\{s_e^{t-1}, a_e^t, s_e^t, r_e^t\}$ in replay buffer for each agent $e \in \mathbb{AGT}$.

Finally, network parameters of both the actor and critic networks within each agent should be updated at a certain frequency for achieving the global optima. For each agent $e \in \mathbb{AGT}$, we employ the PPO method [16] in network updating, to fully explore the time-varying environment states and efficiently exploit the historical knowledge. Specifically, the PPO-based updating process includes (i) calculating the rewards-to-go of e, i.e., the discounted rewards

$$J_e(t) = \sum\nolimits_{t=1}^{T} \eta^t r_e^t; \tag{3.45}$$

(ii) calculating e's advantage function

$$A_e(t) = J_e(t) - V_{\phi_e}(s_e^t); \tag{3.46}$$

(iii) calculating the loss function of e's actor network:

$$L_{\theta_e} = \sum\nolimits_{t=1}^{T} A_e(t) \min\left(\frac{p_{\theta_e}(a_e^t|s_e^t)}{p_{\theta_e'}(a_e^t|s_e^t)}, clip\left(\frac{p_{\theta_e}(a_e^t|s_e^t)}{p_{\theta_e'}(a_e^t|s_e^t)}, 1-\epsilon, 1+\epsilon\right)\right), \tag{3.47}$$

where $p_{\theta_e}(a_e^t|s_e^t)$ is the probability of the e's actor network with parameter θ_e choosing action a_e^t at state s_e^t, θ_e' is the parameter of the e's actor network before updating, and $clip(\cdot)$ is the clip function with predefined hyperparameter ϵ; and (iv) calculating the loss function of the e's critic network:

$$L_{\phi_e} = \sum\nolimits_{t=1}^{T} (V_{\phi_e}(s_e^t) - J_e(t))^2. \tag{3.48}$$

Then, the network parameters θ_e and ϕ_e for each agent $e \in \mathbb{AGT}$ can be updated for minimizing their corresponding loss functions L_{θ_e} and L_{ϕ_e} via random gradient decedent methods, e.g. Adam optimizer.

In summary, all detailed steps of the proposed DMO approach are presented in Algorithm 2.

Theorem 3.3 (Computational complexity of DMO) *The computational complexity of the proposed DMO approach is* $O\big(\boldsymbol{I}\big(2T(|\mathcal{S}_{PL}| + |\mathcal{A}_{PL}|) + |\mathcal{B}|T(|\mathcal{S}_R| + |\mathcal{A}_R|) + 6K_{hid}^2 + |\mathcal{C}||\mathcal{B}| + \varpi|\mathcal{N}|L_n(|\mathcal{B}| - L_n)\big)\big)$*, where* $\boldsymbol{I}$ *is the maximum number of training steps, and* ϖ *denotes the maximum number of times to call the SOCF algorithm within each training step.*

Proof The computational complexity of the DMO approach primarily arises from the PPO-based training process integrated with the execution of GS and SOCF algorithms.

In each training step, the complexity of a forward pass through the actor or critic network of each agent with 3 fully connected layers is $O(3K_{hid}^2)$, where K_{hid}^2 denotes the number of hidden neurons in each layer. Meanwhile, each training step also involves a back propagation with the same computational complexity $O(3K_{hid}^2)$. Meanwhile, updating network parameters requires computing the PPO loss, which involves evaluating the value functions for all states and actions of two agents in MDP $\mathcal{M}_{PL}$ and $|\mathcal{B}|$ agents in MDP $\mathcal{M}_R$ across all time frames. Thus, the computational complexity is $O\big(2T(|\mathcal{S}_{PL}| + |\mathcal{A}_{PL}|) + |\mathcal{B}|T(|\mathcal{S}_R| + |\mathcal{A}_R|)\big)$.

It has been widely proved that the computational complexity of the well-known GS algorithm is $O\big(|\mathcal{C}||\mathcal{B}|\big)$. Besides, in SOCF algorithm, the number of coalitions that a sensor can simultaneously join is at most L_n, resulting in at most $L_n(|\mathcal{B}| - L_n)$ attempts of transfer operations. Therefore, the computational complexity of the SOCF algorithm is $O\big(|\mathcal{N}|L_n(|\mathcal{B}| - L_n)\big)$.

To sum up, the computation complexity of DMO can be expressed as $O\big(\mathbf{I}\big(2T(|\mathcal{S}_{PL}| + |\mathcal{A}_{PL}|) + |\mathcal{B}|T(|\mathcal{S}_R| + |\mathcal{A}_R|) + 6K_{hid}^2 + |\mathcal{C}||\mathcal{B}| + \varpi|\mathcal{N}|L_n(|\mathcal{B}| - L_n)\big)\big)$.

3.5 Simulation Results

In this section, simulations are conducted to evaluate the performance of the proposed DMO approach in optimizing the long-term system performance for the federated DT construction. All simulation results are obtained by taking average over 1000 runs with various parameter settings.

Algorithm 2: DMO Approach for $(\Pi_t^{Up*}, \Pi_t^{Low*})_{\forall t}$

Input: Locations of all devices, importance weights $\{I_c(t)\}_{\forall c}$.
Output: The optimal strategies in $\mathcal{G}^H$, i.e., $(\Pi_t^{Up*}, \Pi_t^{Low*})$.

Randomly initialize: Parameters of both the actor and critic networks within all agents, all decision variables and $\mathcal{CP}(1)$;
for *training step* $= 1, 2, ..., $ *max steps* **do**
  for $t = 1, 2, ..., T$ **do**
    $\mathcal{CP}(t) \leftarrow \mathcal{CP}^*(t-1)$;
    Calculate $s_b^t, \forall b \in \mathcal{B}$;
    for *Given all possible condition of* $\{x_{c,b}(t)\}_{\forall c,b}$ **do**
      while $\mathcal{CP}^*(t)$ *is not fixed* **do**
        Each agent $agt_b, \forall b \in \mathcal{B}$ generates action $\pi_b^R(t)$ based on s_b^t, respectively;
        Call SOCF Algorithm to get $\mathcal{CP}^*(t)$;
        $\mathcal{CP}(t) \leftarrow \mathcal{CP}^*(t)$;
    Calculate s_C^t and $s_{\mathcal{B}}^t$;
    agt_{CS} and $agt_{\mathcal{B}}$ generate actions $\{\mathbf{p}_c(t)\}_{\forall c}$ and $\{\mathbf{p}_b(t)\}_{\forall b}$ based on s_{CS}^t and $s_{\mathcal{B}}^t$, respectively;
    Call GS algorithm to calculate $\{x_{c,b}(t)\}_{\forall c,b}$;
    Calculate the reward r_e^t of each agent $e \in \mathbb{AGT}$;
    Update the state s_e^{t+1} of each agent $e \in \mathbb{AGT}$;
    for *each agent* $e \in \mathbb{AGT}$ **do**
      Store tuple $\{s_e^t, a_e^t, r_e^t, s_e^{t+1}\}$ in e's replay buffer;
  if *training step* % *update steps* $= 0$ **then**
    for *each agent* $e \in \mathbb{AGT}$ **do**
      Calculate e's rewards-to-go $J_e(t) = \sum_{t=1}^{T} \eta^t r_e^t$;
      Calculate e's advantage $A_e(t) = J_e(t) - V_{\phi_e}(s_e^t)$;
      Generate the loss for e's actor network:
      $L_{\theta_e} = \sum_{t=1}^{T} A_e(t) \min(\frac{p_{\theta_e}(a_e^t|s_e^t)}{p_{\theta_e'}(a_e^t|s_e^t)}, clip(\frac{p_{\theta_e}(a_e^t|s_e^t)}{p_{\theta_e'}(a_e^t|s_e^t)}, 1-\epsilon, 1+\epsilon))$;
      Generate the loss for e's critic network: $L_{\phi_e} = \sum_{t=1}^{T} (V_{\phi_e}(s_e^t) - J_e(t))^2$;
      Update θ_e and ϕ_e via Adam Optimizer.

3.5.1 Simulation Settings

We consider a federated DT construction framework enabled by distributed sensing within a $1000\,\text{m} \times 1000\,\text{m}$ geographic area, where $|\mathcal{N}| = 20$ sensors and $|\mathcal{B}| = 5$ ESs are randomly scattered, aiming to build a global DT model consisting of $|C| = 5$ partial-DTs. All key simulation parameters are listed in Table 3.2, most of which are set according to conventional configurations in the literature [4, 5, 19, 26, 44, 49, 55, 59, 60]. Besides, the structure of actor and critic networks, along with hyperparameter settings for the proposed DMO approach, are provided in Table 3.3. Note that some parameters may vary for different evaluation purposes.

Table 3.2 Simulation parameters

Parameter	Value	Parameter	Value
$I_c(t)$	[0, 1]	L_n	3
W	[1, 5] MHz	$\lvert\mathcal{W}\rvert$	10
β	200	ρ_b	10^{-16}
L, δ, γ	{8, 0.02, 2}	F_b, F_{CS}	{64, 3000} MHz
$R^T_{CS,b}, R^T_{b,CS}, R^T_{b,b'}$	[1, 3] Mbps	Υ_b	120 cycles/bit
R^T_{Min}	[1, 5] kbps	$d_{n,c}(t)$	[200, 600] kbits
D_c	[1, 5] Mbits	P_b, P_n	[5, 33] dBm
$P^{Max}_{CS}, P^{Idle}_{CS}, P^{Leak}_{CS}$	[1, 60] Watt	n_b^2	−104 dbm/Hz
ξ, κ, C^{Conf}	[0.1, 15]	δ_T	7.6s
$H_{n,b,w}(t)$	{0.2, 0.4, 0.6}	η	0.92

Table 3.3 Neural network settings

Actor network		
Linear layers with size $\{(K_e^I, 64), (64, 64), (64, K_e^O)\}$		
Critic network		
Linear layers with size $\{(K_e^I, 64), (64, 64), (64, 1)\}$		
Agent e	Input size K_e^I	Output size K_e^O
$agt_{\mathcal{C}}$	$3\lvert\mathcal{C}\rvert + \lvert\mathcal{B}\rvert\big(1 + \lvert\mathcal{N}\rvert(\lvert\mathcal{W}\rvert + 1)\big)$	$\lvert\mathcal{B}\rvert\lvert\mathcal{C}\rvert$
$agt_{\mathcal{B}}$	$\lvert\mathcal{C}\rvert(4 + \lvert\mathcal{B}\rvert)$	$\lvert\mathcal{B}\rvert\lvert\mathcal{C}\rvert$
agt_b	$(4 + \lvert\mathcal{B}\rvert) + \lvert\mathcal{B}\rvert\lvert\mathcal{N}\rvert$	$\lvert\mathcal{N}\rvert + 1$

Furthermore, to show the effectiveness of the proposed federated DT construction framework, several alternative DT construction schemes are simulated for comparison.

- *Centra* [31]: All sensing data is directly uploaded from sensors to the central cloud for constructing the global DT model.
- Non-overlap [38]: A similar edge-cloud collaborative federated DT construction framework is utilized while each sensor is associated with only one ES in each time frame, disabling potential overlaps.

Additionally, to evaluate the superiority of the proposed DMO approach in optimizing the long-term performance for our considered federated DT construction framework, the following algorithms are simulated as benchmarks.

- *GRE*: All optimization decisions are greedily determined to maximize the utility function $\mathcal{U}(t)$ within each single time frame, ignoring the interdependence between different time frames.

- *QL* [53]: All optimization decisions are determined to maximize state-action values (i.e., Q-values) frame-by-frame, which are stored and updated by predefined tables rather than deep neural networks.
- *LBCD* [11]: All optimization decisions are relaxed to continuous form and dynamically determined by block coordinate descent algorithm, by approximately decomposing the original problem by Lyapunov method into instant problem in each time frame.
- *PTS* [41]: All optimization decisions are determined using preference gravity-based tabu search algorithm within each single time frame, except that ES-sensor associations are optimized frame-by-frame.
- *MAB* [55]: All optimization decisions are made dynamically by the contextual multi-armed bandit algorithm (MAB), where environment states serve as contexts to guide the long-term decision-makings by MAB-based agents.

3.5.2 Performance Evaluations

Figure 3.4 illustrates the convergence of the proposed DMO approach. To assess the long-term performance, we define the cumulative system utility as $\sum_{t=0}^{T} \mathcal{U}(t) - C^{Conf}\mathcal{F}(t) - \max\left\{\psi\left(\tau^{Total}(t) - \delta_T\right), 0\right\}$, where T is set to 100 time frames. The results show that the system utility converges rapidly, indicating that all strategies can quickly converge to the equilibrium in the formulated hierarchical game $\mathcal{G}^H$. This is because of not only the use of virtual agents in the DRL, which reduce the complexity of both state and action spaces to enhance training efficiency, but also the adoption of PPO in the strategy exploration, which accelerates the convergence. All these findings align with the theoretical analysis presented in Theorem 3.1 and demonstrate the efficiency of the proposed DMO approach.

Figure 3.5 examines the convergence of the designed SOCF algorithm, showing that the total utility of all ESs, i.e., $\sum_{b\in\mathcal{B}} U_b(t)$ in each time frame t, can converge within a limited number of iterations, and the cumulative sum of all ESs' utilities $\sum_{t'=1}^{t}\sum_{b\in\mathcal{B}} U_b(t')$ increases as t increases. This indicates that the equilibrium of the overlapping coalition formation subgame $\mathcal{G}_t^{Low}$ can be reached within each time frame. The reason is that no sensor has the motivation to deviate from its coalition once the final coalition partition, $\mathcal{CP}^*(t)$, is achieved when the iteration of SOCF algorithm terminates, which well matches the theoretical analyses of Lemma 3.2. Moreover, it is evident that the total number of iterations required to reach $\mathcal{CP}^*(t)$ decreases as t increases. This is due to the diminishing marginal benefit of additional distributed sensing data in improving partial-DT model quality. As a result, when a substantial amount of historical data is available, ESs are less likely to associate with additional sensors (i.e., being reluctant to make frequent changes of the coalition partition), leading to a reduction in iterations of SOCF algorithm.

Figure 3.6 demonstrates the performance of federated DT construction using the proposed DMO approach, and show how long-term system performance $\mathcal{U}_{Sys}$, is

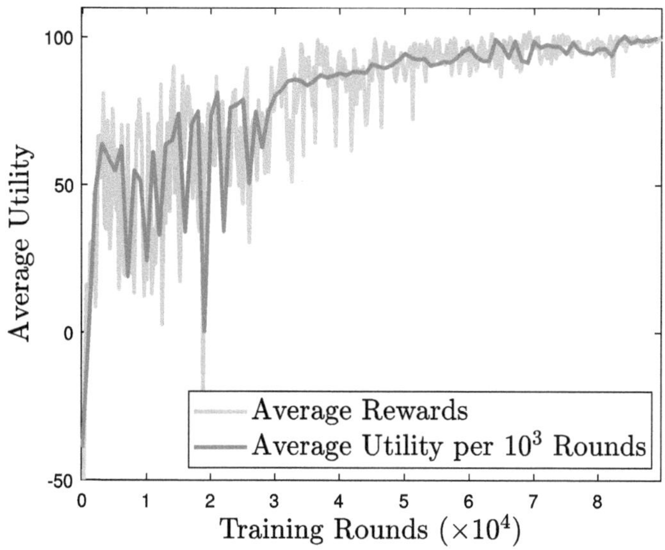

Fig. 3.4 Convergence of the proposed DMO approach

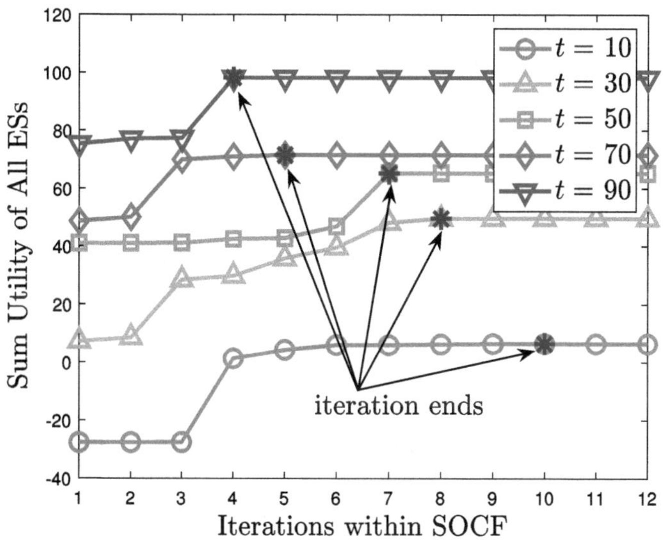

Fig. 3.5 Convergence of the designed SOCF algorithm

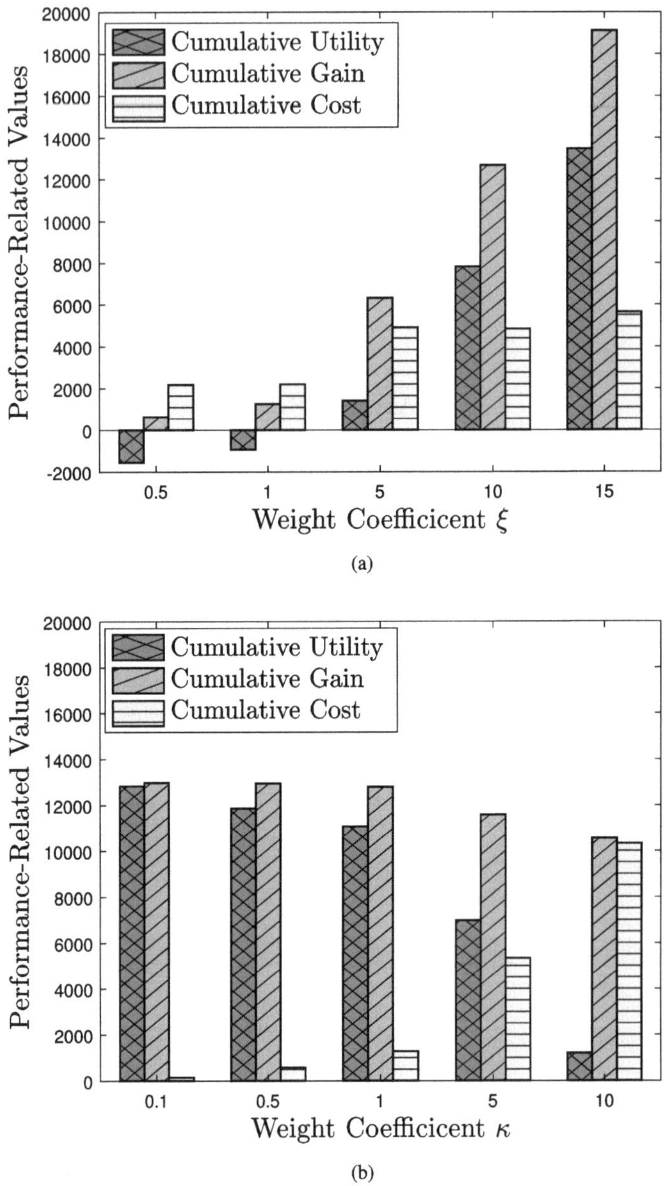

Fig. 3.6 Long-term performance with different ξ and κ

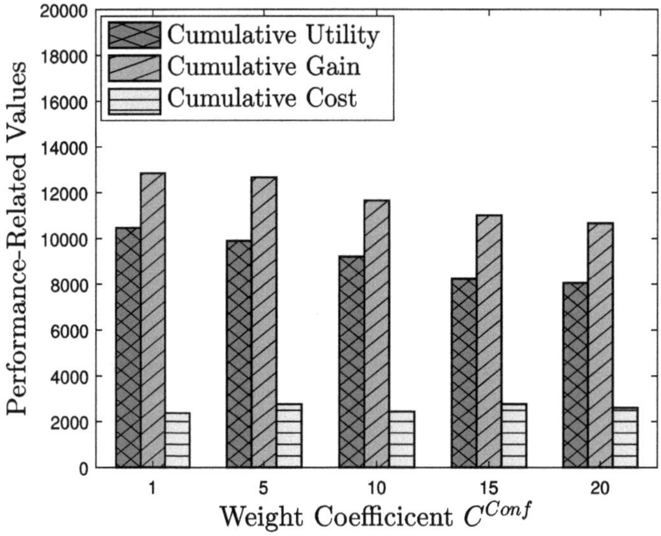

Fig. 3.7 Long-term performance with different C^{Conf}

influenced by weight coefficients ξ and κ. In Fig. 3.6a, it is observed that as ξ increases, both the cumulative utility and gain rise, while the cumulative cost tends to be stable. This occurs because, with a larger ξ, associating more sensors, i.e., collecting more feature data, significantly improves the quality of the global DT model. However, since the number of sensors that can be associated with each ES is limited, the cumulative cost is constrained. In Fig. 3.6b, while the cumulative cost sharply increases with κ, the DMO approach can still produce a relatively large system utility. This is because the DMO approach dynamically associates sensors with lower costs (e.g., communication overheads and energy consumptions) in data collection. Additionally, the cumulative gain in Fig. 3.6b is noticeably lower than that in Fig. 3.6a because, as κ becomes sufficiently large, sensor associations are limited by the overwhelming cost, thereby restricting potential gains.

Figure 3.7 illustrates the impact of configuration cost C^{Conf} on system's cumulative utility when the proposed DMO approach is employed. It can be seen that, as C^{Conf} increases, the cumulative gain gradually decreases. This is because, when C^{Conf} becomes sufficiently large, ESs are less willing to associate with additional sensors, i.e., dynamic sensor association $y_{b,n}(t)$ is constrained to a fixed one. This restriction limits the amount of feature data that each ES can acquire, and thus deter further improvements in model quality. Moreover, we can observe that the cumulative cost does not increase drastically. This is because the designed SOCF algorithm can result in less frequent coalition changes for each ES if C^{Conf} is considerably large.

Figure 3.8 demonstrates the effectiveness of the proposed federated DT construction framework over other DT construction schemes. It can be observed from Fig. 3.8a

Fig. 3.8 Performance comparison on different DT construction frameworks. **a** Comparison on cumulative utility, **b** Comparison on cumulative gain, **c** Comparison on cumulative cost

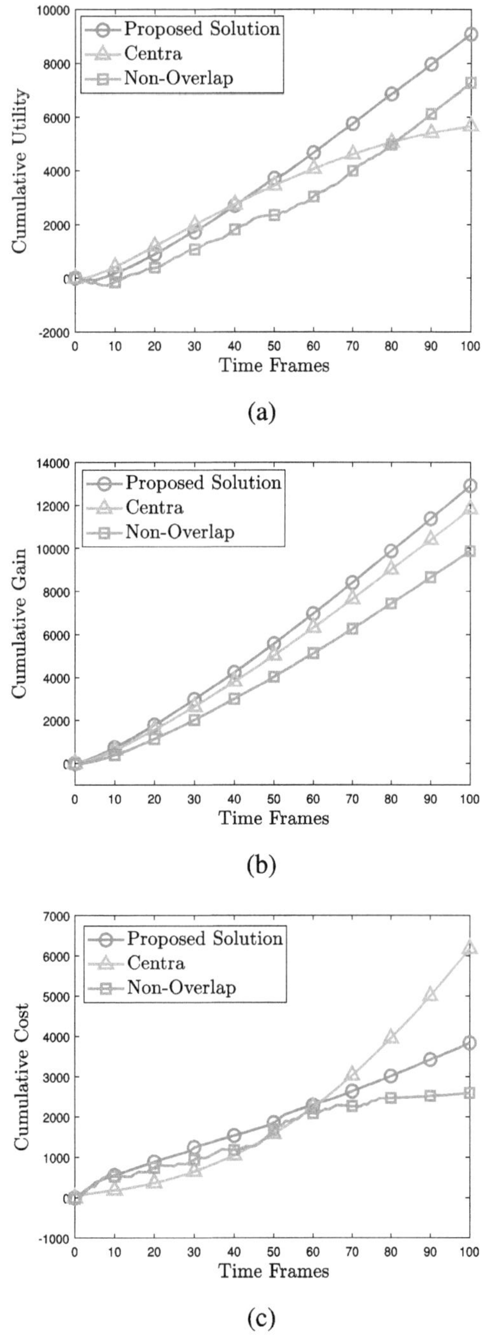

and b that the proposed framework outperforms both *Centra* and *Non-overlap* in terms of the cumulative utility and gain. This is because the proposed framework allows sensors to be associated in an overlapped manner, enabling more efficient feature data collection, and thus results in a higher quality of the global DT model. Besides, although *Centra* allows sensors to directly upload their data to the cloud, the number of sensors that can be associated is limited by the available subcarriers and strict delay constraints, leading to a lower cumulative gain. Figure 3.8c shows that, the proposed framework achieves lower cumulative costs than *Centra* and only slightly higher costs than *Non-overlap*. This is due to the fact that (i) in *Centra*, the centralized data collection and DT creation introduce significant communication and computation latency, and (ii) in *Non-overlap*, while associating sensors without overlaps reduces the energy consumption for data transmission, compared with partial-DT creations and global DT integration that are relatively more resource-intensive, this actually accounts for only a small portion of the total energy consumption, which can hardly offset the loss of cumulative gain.

Figure 3.9 compares the proposed DMO approach with existing optimizing methods in terms of the long-term system performance. We can see that (i) in Fig. 3.9a, the proposed approach outperforms all benchmarks in terms of the cumulative utility, while *GRE* exhibits the worst performance; (ii) in Fig. 3.9b, the cumulative gain of the proposed solution approach is significantly lower than that of all other benchmarks, while *QL* exhibits the worst gain; and (iii) in Fig. 3.9c, the cumulative cost of the proposed solution approach is significantly lower than that of all other benchmarks except *LBCD*. These is because (i) although *GRE* can achieve a relatively high cumulative gain, it suffers from excessive costs due to greedily prioritizing the ES with the most feature data to create the partial-DT with the highest importance, ignoring the resulted energy consumptions and communication overheads; (ii) *LBCD* requires relaxation of originally discrete decision variables and an approximate decomposition of the long-term objective function, leading to a near-optimal solution when facing unpredictable system dynamics in federated DT construction; (iii) although the proposed DMO approach and *QL* are all based on reinforcement learning, they achieve different performance because the proposed DMO approach can better approximate the long-term non-convex cumulative utility through the neural network based actor and critic networks, rather than a predefined table in *QL*; (iv) the proposed DMO approach have stronger exploration capabilities compared to heuristic algorithms (i.e., *PTS*) in interacting with the dynamic environment, reducing the possibility of falling into the local optima; (v) unlike *MAB*, the proposed approach can fully leverage historical training data through the replay buffer of each agent, enhancing the long-term system performance.

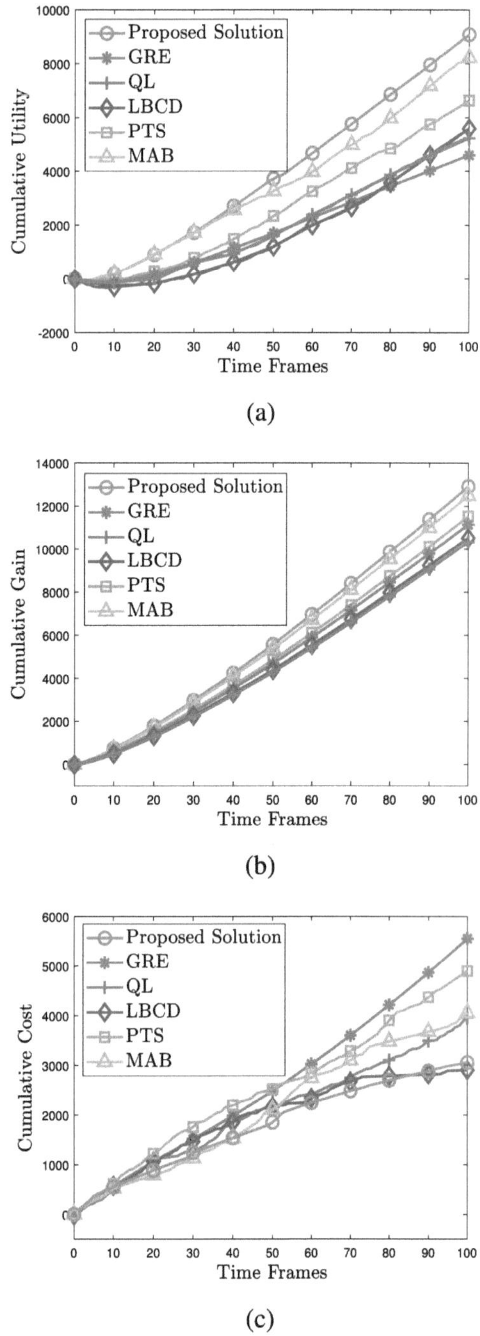

Fig. 3.9 Performance comparison on different optimization methods. **a** Comparison on cumulative utility, **b** Comparison on cumulative gain, **c** Comparison on cumulative cost

3.6 Summary of the Chapter

In this chapter, we have studied a federated DT construction framework enabled by distributed sensing under edge-cloud collaboration. We have formulated an online optimization problem to jointly and dynamically determine partial-DT assignments, ES-sensor associations, and as well as computation and communication resource allocations, for maximizing the long-term system performance. To tackle this complicated problem, we have conducted a transformation, forming an equivalent two-layer hierarchical game, consisting of an upper-layer two-sided matching game and a lower-layer overlapping coalition formation game. After analyzing these games in detail, we have applied the GS algorithm and particularly developed an SOCF algorithm to respectively derive their short-term equilibria, and then proposed a DMO approach to accommodate dynamic settings, thereby solving the original problem. Simulation results show the effectiveness of the federated DT construction framework and demonstrate the superiority of the proposed DMO approach over counterparts.

References

1. Adreani, L., Colombo, C., Fanfani, M., Nesi, P., Pantaleo, G., Pisanu, R.: Rendering 3D city for smart city digital twin. In: Proceedings of IEEE SMARTCOMP, pp. 183–185 (2022). https://doi.org/10.1109/SMARTCOMP55677.2022.00046
2. Altahir, A.A., Asirvadam, V.S., et al.: Optimizing visual sensor coverage overlaps for multiview surveillance systems. IEEE Sens. J. **18**(11), 4544–4552 (2018). https://doi.org/10.1109/JSEN.2018.2825781
3. Brown, K.L.: Stochastic Games and Bayesian Games. University of British Columbia (2020). https://www.cs.ubc.ca/kevinlb/teaching/cs532l%20-%202011-12/lectures/lect10.pdf
4. Cai, J., Shen, X., Mark, J.W., Alfa, A.S.: Semi-distributed user relaying algorithm for amplify-and-forward wireless relay networks. IEEE Trans. Wirel. Commun. **7**(4), 1348–1357 (2008)
5. Cao, H., Cai, J.: Distributed multiuser computation offloading for cloudlet-based mobile cloud computing: a game-theoretic machine learning approach. IEEE Trans. Veh. Technol. **67**(1), 752–764 (2018)
6. Carbontribe: Cardat collection on OpenSea (Testnet) (2024). https://testnets.opensea.io/collection/cardat
7. Chen, C., Liu, J.X., Shi, J.W., Li, Q.W.: Influence on flexibility of networked C2 structure with different importance of tasks. In: 2016 13th International Computer Conference on Wavelet Active Media Technology and Information Processing (ICCWAMTIP), pp. 473–476 (2016). https://doi.org/10.1109/ICCWAMTIP.2016.8079897
8. Chen, J., Shi, Y., Yi, C., Du, H., Kang, J., Niyato, D.: Generative AI-driven human digital twin in IoT-healthcare: a comprehensive survey. IEEE Internet Things J. **11**(21), 34749–34773 (2024)
9. Chen, J., Wu, Q., et al.: Joint task assignment and spectrum allocation in heterogeneous UAV communication networks: a coalition formation game-theoretic approach. IEEE Trans. Wirel. Commun. **20**(1), 440–452 (2021). https://doi.org/10.1109/TWC.2020.3025316
10. Chen, J., Yi, C., Du, H., Niyato, D., Kang, J., Cai, J., Shen, X.: A revolution of personalized healthcare: enabling human digital twin with mobile AIGC. IEEE Netw. **38**(6), 234–242 (2024)
11. Chen, J., Yi, C., Okegbile, S.D., Cai, J., Shen, X.S.: Networking architecture and key supporting technologies for human digital twin in personalized healthcare: a comprehensive survey. IEEE Commun. Surv. Tutor. **26**(1), 706–746 (2024)

12. Chen, J., Yi, C., Wang, R., Zhu, K., Cai, J.: Learning aided joint sensor activation and mobile charging vehicle scheduling for energy-efficient WRSN-based industrial IoT. IEEE Trans. Veh. Technol. **72**(4), 5064–5078 (2023)
13. Chen, R., Yi, C.: A game-theoretic online optimization for federated digital twin construction via wireless sensing. In: Proceedings of IEEE ICC, Montreal, Canada (2025)
14. Chen, R., Yi, C., Zhou, F., Kang, J., Wu, Y., Niyato, D.: Federated digital twin construction via distributed sensing: a game-theoretic online optimization with overlapping coalitions. IEEE Trans. Mob. Comput. pp. 1–17 (2025)
15. Chen, R., Yi, C., Zhu, K., Cai, J., Chen, B.: A DRL-based hierarchical game for physical layer security with dynamic trilateral coalitions. In: Proceedings of IEEE ICC, pp. 4495–4500 (2023)
16. Chen, R., Yi, C., Zhu, K., Chen, B., Cai, J., Guizani, M.: A three-party hierarchical game for physical layer security aware wireless communications with dynamic trilateral coalitions. IEEE Trans. Wirel. Commun. **23**(5), 4815–4829 (2024)
17. Chen, W., Zhao, S., et al.: Generalized user grouping in NOMA based on overlapping coalition formation game. IEEE J. Sel. Areas Commun. **39**(4), 969–981 (2021). https://doi.org/10.1109/JSAC.2020.3018832
18. Dewanto, V., Gallagher, M.: Examining average and discounted reward optimality criteria in reinforcement learning. In: Proceedings of AI 2022: Advances in Artificial Intelligence, pp. 800–813. Springer (2022)
19. Di, B., Wang, T., Song, L., Han, Z.: Collaborative smartphone sensing using overlapping coalition formation games. IEEE Trans. Mob. Comput. **16**(1), 30–43 (2017). https://doi.org/10.1109/TMC.2016.2538224
20. Gupta, D., Moni, S.S., Tosun, A.S.: Integration of digital twin and federated learning for securing vehicular Internet of Things. In: Proceedings of the International Conference on Research in Adaptive and Convergent Systems, pp. 1–8 (2023)
21. Han, Z., Niyato, D., Saad, W., Başar, T., Hjørungnes, A.: Game Theory in Wireless and Communication Networks: Theory, Models, and Applications. Cambridge University Press, Cambridge, UK (2012)
22. He, X., Lv, C.: Robotic control in adversarial and sparse reward environments: a robust goal-conditioned reinforcement learning approach. IEEE Trans. Artif. Intell. **5**(1), 244–253 (2024)
23. Jia, W., Wang, W., et al.: From simple digital twin to complex digital twin part I: a novel modeling method for multi-scale and multi-scenario digital twin. Adv. Eng. Inform. **53**, 101706 (2022)
24. Jiang, F., Ma, L., Broyd, T., Chen, K.: Digital twin and its implementations in the civil engineering sector. Autom. Constr. **130**, 103838 (2021)
25. Kobayashi, M., Uekumasu, T., Saruwatari, S., Watanabe, T.: Dynamic channel assignment for downlink and uplink decoupling in wireless networks. IEEE Access **11**, 38631–38642 (2023). https://doi.org/10.1109/ACCESS.2023.3267090
26. Li, G., Cai, J.: An online incentive mechanism for collaborative task offloading in mobile edge computing. IEEE Trans. Wirel. Commun. **19**(1), 624–636 (2020)
27. Li, J., Guo, S., et al.: Digital twin-enabled service provisioning in edge computing via continual learning. IEEE Trans. Mobile Comput. **23**(6), 7335–7350 (2024)
28. Li, X., Wei, J., Wang, H., Dong, L., Chen, R., Yi, C., Cai, J., Niyato, D., Shen, X.: Towards intelligent transportation with pedestrians and vehicles in-the-loop: a surveillance video-assisted federated digital twin framework. IEEE Netw. pp. 1–1 (2025)
29. Liu, D., Chen, R., Yi, C., Zhang, T., Li, X., Li, J., Wang, R., Zhu, K.: A hierarchical game for physical layer security aware cooperative communications with dynamic interchangeable relaying and jamming. IEEE Trans. Veh. Technol. **74**(1), 968–983 (2025)
30. Liu, D., Chen, R., Zhang, T., Yi, C.: A DRL-based hierarchical game for physical layer security aware cooperative communications. In: Proceedings of IEEE ICCC, pp. 1769–1774 (2024)
31. Liu, Q., Han, T., Xie, J.L., Kim, B.: Livemap: real-time dynamic map in automotive edge computing. In: Proceedings of IEEE INFOCOM, pp. 1–10 (2021)
32. Lu, Q., Parlikad, A.K., et al.: Developing a digital twin at building and city levels: case study of west Cambridge campus. J. Manage. Eng. **36**(3) (2020)

33. Man, T., Zhu, H., Chen, R., Yi, C.: Hierarchical DRL-based multi-motor control with torque synchronization in industrial IoT. In: International Conference on Wireless Artificial Intelligent Computing Systems and Applications (WASA), pp. 11–21. Springer (2025)
34. Microsoft: Azure Digital Twins documentation (2025). https://learn.microsoft.com/en-us/azure/digital-twins/
35. Mihai, S., Yaqoob, M., et al.: Digital twins: a survey on enabling technologies, challenges, trends and future prospects. IEEE Commun. Surv. Tutor. **24**(4), 2255–2291 (2022)
36. Mukhametzyanov, I.Z.: Non-linear multivariate normalization methods. In: Normalization of Multidimensional Data for Multi-criteria Decision Making Problems: Inversion, Displacement, Asymmetry, pp. 167–194. Springer (2023)
37. Nayyar, A., Gupta, A., et al.: Common information based Markov perfect equilibria for stochastic games with asymmetric information: finite games. IEEE Trans. Autom. Control **59**(3), 555–570 (2014)
38. Ng, J.S., et al.: Reputation-aware hedonic coalition formation for efficient serverless hierarchical federated learning. IEEE Trans. Parallel Distrib. Syst. **33**(11), 2675–2686 (2022)
39. Okegbile, S.D., Cai, J., Yi, C., Niyato, D.: Human digital twin for personalized healthcare: vision, architecture and future directions. IEEE Netw. **37**(2), 262–269 (2023)
40. Pham, Q.V., Le, M., et al.: Energy-efficient federated learning over UAV-enabled wireless powered communications. IEEE Trans. Veh. Technol. **71**(5), 4977–4990 (2022). https://doi.org/10.1109/TVT.2022.3150004
41. Qi, N., Huang, Z., et al.: A task-driven sequential overlapping coalition formation game for resource allocation in heterogeneous UAV networks. IEEE Trans. Mob. Comput. **22**(8), 4439–4455 (2023). https://doi.org/10.1109/TMC.2022.3165965
42. Qian, Z., Chen, R., Yi, C., Zhai, X., Chen, B.: Collision avoidance control for autonomous driving with multiple dynamic obstacles in IoV: a prediction-enhanced APF-based approach. IEEE Internet Things J. (2025)
43. Rayhana, R., Bai, L., Xiao, G., Liao, M., Liu, Z.: Digital twin models: functions, challenges, and industry applications. IEEE J. Radio Freq. Identif. **8**, 282–321 (2024)
44. Shang, S., Yi, C., Zhang, T., Chen, R., Cai, J.: Latency-energy aware dynamic application placement for edge computing: a vacation queue based optimization approach. IEEE Trans. Netw. Sci. Eng **11**(2), 2249–2263 (2024)
45. Shi, S., Hu, C., Wang, D., Zhu, Y., Han, Z.: Federated HD map updating through overlapping coalition formation game. IEEE Trans. Mobile Comput. **23**(2), 1641–1654 (2024). https://doi.org/10.1109/TMC.2023.3241090
46. Shi, Y., Tong, Y., et al.: Efficient approximate range aggregation over large-scale spatial data federation. IEEE Trans. Knowl. Data Eng. **35**(1), 418–430 (2023). https://doi.org/10.1109/TKDE.2021.3084141
47. Shi, Y., Wang, J., Yi, C., Wang, R., Chen, B.: Proactive application deployment for MEC: An adaptive optimization based on imperfect multi-dimensional prediction. IEEE Trans. Veh, Technol (2025)
48. Shi, Y., Yang, Y., Yi, C., Chen, B., Cai, J.: Towards online reliability-enhanced microservice deployment with layer sharing in edge computing. IEEE Internet Things J. **11**(13), 23370–23383 (2024)
49. Shi, Y., Yi, C., Wang, R., Wu, Q., Chen, B., Cai, J.: Service migration or task rerouting: a two-timescale online resource optimization for MEC. IEEE Trans. Wirel. Commun. **23**(2), 1503–1519 (2024)
50. Sun, W., Xu, N., Wang, L., Zhang, H., Zhang, Y.: Dynamic digital twin and federated learning with incentives for air-ground networks. IEEE Trans. Netw. Sci. Eng. **9**(1), 321–333 (2022). https://doi.org/10.1109/TNSE.2020.3048137
51. Wang, Y., Su, Z., Pan, Y., Luan, T.H., Li, R., Yu, S.: Social-aware clustered federated learning with customized privacy preservation. IEEE/ACM Trans. Netw. **32**(5), 3654–3668 (2024)
52. Wu, J., Chen, R., Yi, C.: A DRL-based deviation-aware federated digital twin construction over wireless edge network. In: International Conference on Wireless Artificial Intelligent Computing Systems and Applications (WASA), pp. 196–208. Springer (2025)

53. Xu, H., Wu, J., Pan, H., Gu, J., Guan, X.: Delay safety-aware digital twin empowered industrial sensing-actuation systems using transferable and reinforced learning. IEEE Trans. Ind. Inform. **19**(11), 10811–10820 (2023)
54. Xu, L., Yu, H., et al.: Digital twin for aquaponics factory: analysis, opportunities, and research challenges. IEEE Trans. Ind. Inform. **20**(4), 5060–5073 (2024)
55. Xu, Z., Li, D., et al.: Energy or accuracy? Near-optimal user selection and aggregator placement for federated learning in MEC. IEEE Trans. Mobile Comput. **23**(3), 2470–2485 (2024). https://doi.org/10.1109/TMC.2023.3262829
56. Yang, X., Zheng, J., Luan, T.H., Li, R., Su, Z., Dong, M.: Data synchronization for vehicular digital twin network. In: Proceedings of IEEE GLOBECOM, pp. 5795–5800 (2022). https://doi.org/10.1109/GLOBECOM48099.2022.10000831
57. Yang, Y., Shi, Y., Chen, R., Yi, C., Kang, J.: Online optimization of edge empowered human digital twin deployment and task offloading. In: Proceedings of IEEE ICCC, pp. 856–861 (2024)
58. Yang, Z., Chen, M., et al.: Energy efficient federated learning over wireless communication networks. IEEE Trans. Wirel. Commun. **20**(3), 1935–1949 (2021). https://doi.org/10.1109/TWC.2020.3037554
59. Yi, C., Cai, J., Su, Z.: A multi-user mobile computation offloading and transmission scheduling mechanism for delay-sensitive applications. IEEE Trans. Mob. Comput. **19**(1), 29–43 (2020)
60. Yi, C., Huang, S., Cai, J.: An incentive mechanism integrating joint power, channel and link management for social-aware D2D content sharing and proactive caching. IEEE Trans. Mobile Comput. **17**(4), 789–802 (2018)
61. Yu, X., Yang, Y., Chen, R., Yi, C.: Online optimization of edge vehicle digital twin migration with adaptive mobility prediction. In: Proceedings of IEEE VTC, Chengdu, China (2025)
62. Zhang, T., Mao, S.: Energy-efficient federated learning with intelligent reflecting surface. IEEE Trans. Green Commun. Netw. **6**(2), 845–858 (2022). https://doi.org/10.1109/TGCN.2021.3126795
63. Zhang, Z., Liu, Y., Peng, Z., Chen, M., Xu, D., Cui, S.: Digital twin-assisted data-driven optimization for reliable edge caching in wireless networks. IEEE J. Sel. Areas Commun. **42**(11), 3306–3320 (2024). https://doi.org/10.1109/JSAC.2024.3431575
64. Zhou, H., Chen, R., Yi, C., Li, J., Cai, J.: A three-party repeated coalition formation game for PLS in wireless communications with IRSS. In: Proceedings of IEEE WCNC, pp. 1–6 (2024)
65. Zhou, H., Chen, R., Yi, C., Zhang, J., Kang, J., Cai, J., Guizani, M.: A repeated coalition formation game for physical layer security aware wireless communications with third-party intelligent reflecting surfaces. IEEE Trans. Wirel. Commun. pp. 1–1 (2025)
66. Zhou, S., et al.: A machine-learning-based surrogate modeling methodology for submodel integration in the holistic railway digital twin platform. Procedia CIRP **119**, 345–350 (2023)
67. Zhu, W., Chen, R., Yi, C., Cai, J.: Edge-assisted video transmission with adaptive key frame selection: a hierarchical DRL approach. In: Proceedings of IEEE Biennial Symposium on Communications, pp. 89–94. IEEE (2023)

Chapter 4
Quality-of-Experience Driven Digital Twin Service Interaction with Uncertain Evolution

4.1 Introduction

Digital twin (DT) is envisioned to be a groundbreaking technology, primarily aiming to create high-fidelity and interactive virtual counterparts of physical entities, known as DT models, which serve as hyper-realistic and vivid avatars for real-time monitoring and predictive analytics, enabling a deeper and forward-looking understanding of the physical world [5, 13, 57]. To leverage the potential of DT services, the immersive interaction with DT models, namely allowing users to access DT models in obtaining insightful feedbacks [29, 60], is extremely important and indispensable [7, 44, 67]. Meanwhile, DT services are shifting from being the cloud-based to ubiquitous edge-based [6, 12, 32, 67], and this urges the need of offering pervasive wireless access to DT models by mobile users anywhere and anytime [54]. In this context, reconfigurable intelligent surface (RIS) [71], due to its powerful capability of flexibly steering signals, mitigating interference and enhancing coverage, is a promising solution for assisting mobile users in DT interactions requiring seamless connectivity under dynamic wireless conditions [8, 15].

Although RIS-assisted wireless communications have been extensively studied in the existing work [22, 47, 52, 70], they commonly apply to general wireless services, while the DT interaction possesses distinct features and requirements, making it fundamentally different. On the one hand, unlike those wireless applications, e.g., edge computing [3, 4, 22, 47] and content sharing [27, 59], which focus on either uplink or downlink communications only, both uplink and downlink play crucial roles in DT interactions. The uplink and downlink in DT interactions are responsible to convey mission-critical while data-intensive user interaction requests and corresponding feedbacks, respectively. For example, when mobile users engage in interactions with educational DT models or healthcare DT models, they first upload their interaction signals, such as motions, voices and facial expressions, via uplink channels. Then, DT models generate feedbacks, such as virtual 3D physical simulation outcomes and multimodal virtual diagnosis responses [5, 51], after receiving the interaction signals,

C. Yi et al., *Self-Evolving Digital Twin Over Wireless Networks*, Wireless Networks,
https://doi.org/10.1007/978-3-032-08120-9_4

and transmit them back to mobile users via downlink channels. On the other hand, in addition to the objective performance metrics, e.g., latency, which are usually concerned in most wireless systems [22, 52, 58]. DT interactions also place significant emphasizes on users' subjective experiences. For instance, besides objective factors, subjective factors, such as the feeling of low authenticity, inconsistency of same tasks and lack of adaptability, may lead to disengagement, increased frustration and reduced learning outcomes in education DT interactions [18], and can even result in misdiagnoses, delayed treatments and life-threatening consequences in healthcare DT interactions [56]. Therefore, to fully enable RIS-assisted DT interactions, it is required to construct a round-trip communication system with equal considerations on uplink and downlink, aiming to maximize a particularly designed quality of experience (QoE) metric composed of both objective and subjective performances. Despite the significance of this issue, it is very challenging for the following reasons.

1. In RIS-assisted DT interactions, mobile users' uplink interaction signal transmissions, the DT model's downlink feedback signal generations and transmissions occur simultaneously, not only sharing the limited communication and computation resources, but also contributing together to end-to-end objective and subjective performances [67, 71]. This indicates that all resource allocations related to RIS-assisted DT interactions, including the phase shift matrix, receive/transmit beamforming matrices in high dimensions, feedback signal rendering resolution and computation resource configurations have to be jointly optimized. Moreover, optimizing the subjective (objective) performance commonly requires enhancing (lowering) feedback signal resolutions, which in turn increases (decreases) communication and computation burdens, ultimately diminishing the objective (subjective) performance. Such an inherent trade-off between objective and subjective performances makes it impossible to address them separately, not to mention that different mobile users may have varying weights towards objective/subjective performances. These motivate us to devise an efficient algorithm for jointly allocating resources in RIS-assisted DT interactions, while balancing objective and subjective performances, taking into account the individual QoE requirement of each mobile user.
2. Naturally, the status of physical entities in the real word is ever-changing and sometimes beyond expectations, which drives uncertain evolutions of the corresponding DT model [43]. Because of such evolutions, the DT scene that mobile users engage in during interactions may continuously evolve. Triggered by this variation and the impacts on mobile users' QoE across different scenes, the resource allocation for RIS-assisted DT interactions becomes scene-specific rather than one-size-fits-all. This requests us to re-solve each scene-specific problem by re-designing appropriate algorithms whenever the DT scene changes. One potential way to circumvent this difficulty is to exploit the scene-specific information under the corresponding DT scene to guide the optimization algorithm to adapt to new changes in emerging scene-specific problems. Generative artificial intelligence (GAI) with the prompt-based learning [2, 49] may be a good fit for achieving this goal, as it can train an algorithm to perform a variety of tasks by

simply modifying the prompt, being highly adaptable and versatile. However, unlike the popular GAI applications, such as question-answer and image generations [2, 48, 49], where the descriptive text-based prompts are commonly adopted, the resource allocation problem for RIS-assisted DT interactions contains complex scene-specific information, which is not in textual form. This necessitates us to further explore the advanced "decision-making trajectory" in the design of prompts, applicable for the considered problem. Such a trajectory is expected to record a sequence of past resource allocation decisions for RIS-assisted DT interactions.

In this chapter, to fill the gap in the literature, we propose a novel GAI-aided approach to tackle the QoE-aware resource allocation for RIS-assisted DT interactions with uncertain evolutions. In the considered system, mobile users interact with a DT model hosted by a DT server deployed on a wireless base station via the assistance of an RIS. Aiming to provide mobile users with maximum QoE in DT interactions, we jointly optimize the phase shift matrix, receive/transmit beamforming matrices, feedback signal rendering resolution and computation resource configurations for uplink and downlink simultaneously, balancing objective and subjective performances. Further taking into account DT model's uncertain evolutions, we formulate a QoE-aware resource allocation problem, consisting of a series of different scene-specific ones corresponding to each DT scene. To obtain a generalized method, we develop a GAI-aided approach, called prompt-guided decision transformer integrated with zero-forcing optimization (PG-ZFO). Specifically, we first reformulate each emerging scene-specific problem into a Markov decision process (MDP). We then design a "decision-making trajectory" based prompt to capture the new scene-specific information, and extend the traditional decision transformer to a prompt-guided decision transformer with strong generalization. Besides, a zero-forcing (ZF)-based optimization algorithm is integrated to help derive high-dimensional decisions, i.e., receive/transmit beamforming matrix, along with the offline training and online execution of PG-ZFO.

The main contributions of this chapter are summarized in the following.

- To the best of our knowledge, we are the first to study the QoE-aware resource allocation for RIS-assisted DT interactions. We focus on distinct features of DT interactions i.e., treating uplink and downlink with equal importance, and emphasizing not only objective performance but also subjective experience. Besides, we consider the uncertain evolutions of the DT model, and formulate the resulted resource allocation problem, which consists of a series of different scene-specific ones.
- We propose a novel GAI-aided approach, called PG-ZFO, which demonstrates strong generalization capabilities to efficiently tackle each scene-specific problem in RIS-assisted DT interactions. When addressing any new scene-specific problem, it only requires modifying the particularly designed prompt, enabling PG-ZFO to adapt to the new problem without the need for re-training.
- We analyze computational complexities of the proposed PG-ZFO's offline training and online execution. Moreover, we conduct extensive simulations to evaluate the

performance of PG-ZFO. Results show that the PG-ZFO not only demonstrates superior generalization capability over different scene-specific problems, but also outperforms counterparts in terms of the sum of all mobile users' QoE in DT interactions.

The rest of this chapter is organized as follows: Sect. 4.2 reviews the related work and highlights the novelties of this chapter. Section 4.3 describes the considered system model and the problem formulation. In Sect. 4.4, a novel GAI-aided approach, i.e., PG-ZFO, is proposed and analyzed. Simulation results are presented in Sect. 4.5, followed by the conclusion in Sect. 4.6.

Symbol: $\mathbb{C}^{M\times N}$ signifies the $M \times N$ dimensional complex space. $\mathbf{I}_M$ stands for M dimensional identity matrix. The transpose and conjugate transpose operator are denoted using $(\cdot)^T$ and $(\cdot)^H$, respectively. Vectors are denoted using lower case bold symbols (e.g., $\mathbf{h}$), and matrices are denoted using upper case bold symbols (e.g., $\mathbf{H}$). The diagonalization, absolute, Euclidean norm are denoted using $diag(\cdot)$, $|\cdot|$, and $||\cdot||$. $\mathcal{CN}(0, 1)$ represents a circularly symmetric complex Gaussian distribution with zero mean and unit variance. $[\cdot]_{mn}$ represent the element at the mth row and nth column at a matrix.

4.2 Related Work

With significant advantages of enhancing the wireless communication performance by providing additional pathways, RIS has been widely applied to various wireless applications. For example, Jiang et al. [22] investigated an RIS-assisted edge computing system, where mobile users were allowed to offload their computing tasks to an edge server through the RIS, targeting to minimize response delay by jointly optimizing communication and computation resources. Similarly, Qi et al. [47] studied the RIS-aided vehicular edge computing, where an RIS was deployed to increase the quality of communication links when vehicles offload computing tasks to nearby road-side edge servers. Besides the above focusing on the uplink, RIS has also been employed to assist downlink communications. Wang et al. [52] explored the RIS-aided unmanned aerial vehicle (UAV) communication, where multiple UAVs were regarded as aerial base stations to serve mobile users with the assistance of RISs. However, all of these work were restricted to either uplink or downlink only, without jointly considering both, potentially assisted by the RIS simultaneously, which is crucial for DT interactions with a particular attention on round-trip communications.

Some researchers have been devoted to the joint optimization of uplink and downlink communications. Name a few, Li et al. [28] jointly optimized bandwidth allocation and power control of uplink and downlink, along with their antenna configuration and subchannel assignment to minimize the average transmit power while ensuring the stringent quality of service (QoS) requirements of wireless applications. Zhang et al. [63] constructed a joint uplink and downlink robust transmission framework in cell-free networks, targeting to maximize the total data throughput in both uplink and downlink. Park et al. [45] proposed a deep reinforcement learning (DRL)-based approach to address the jointly uplink and downlink resource allocation problem, so as to meet the QoS requirements for achievable data rate in both uplink and downlink when mobile users access aerial base stations enabled by UAVs. However, all these work concerned solely on the system's QoS (i.e., objective performance), without considering the subjective experience and their mixed influences on mobile users' QoE, while this is critical in DT interactions yet non-trivial.

In fact, the DT interaction has been recognized as a key issue in both academia and industry. Moreover, several preliminary work have also noticed the significance of optimizing the mixed objective and subjective QoE. For example, Lin et al. [32] devised a communication-efficient DT interaction scheme over edge computing, taking both interaction cost and users' satisfaction degree for measuring mobile users' QoE. Zhang et al. [67] proposed a mobile user and DT model association scheme, aiming to minimize the DT interaction latency while enhancing mobile users' feeling of DT model consistency. However, both these works did not delve deeply into the underlying wireless communications. In addition, although they acknowledge that the DT model is ever-changing and should be dynamically updated, they overlooked the impact of the resulting variations in the DT scene on DT interactions.

In summary, this work differs from the existing literature in the following aspects: (i) we consider the joint uplink and downlink optimization for RIS-assisted DT interactions; (ii) we introduce and maximize a QoE mixing both objective performance and subjective experience; (iii) we study the uncertain evolution of the DT model, addressing the resulted scene-specific resource allocation problem by a novel GAI-aided approach with strong generalization and easy implementation.

4.3 System Model and Problem Formulation

4.3.1 System Overview

We consider an RIS-assisted DT interaction system, as illustrated in Fig. 4.1, consisting of a DT server (denoted as a) deployed on a base station with a uniform linear array (ULA) of M receive/transmit antennas, located at $\mathbf{q_a} = [x_a, y_a, z_a]$, a set of single-antenna mobile users $\mathcal{K}$ requesting DT interaction services with cardinality of $|\mathcal{K}| = K$, and an RIS (denoted as r) with N passive reflecting elements, located at $\mathbf{q_r} = [x_r, y_r, z_r]$. The DT server maintains a high-fidelity and interactive DT model that provides mobile users with engaging DT interaction services. This DT model meticulously mirrors the real-time status of its corresponding physical entity, such as an educational system or a healthcare system in the real world. Mobile users access the DT server by wireless communications, interacting with high-fidelity virtual representations in the DT model. For instance, when mobile users engage in an immersive educational DT, they can interact with virtual teachers built at the DT server whose responses are dynamically aligned with the current status of their real-world coun-

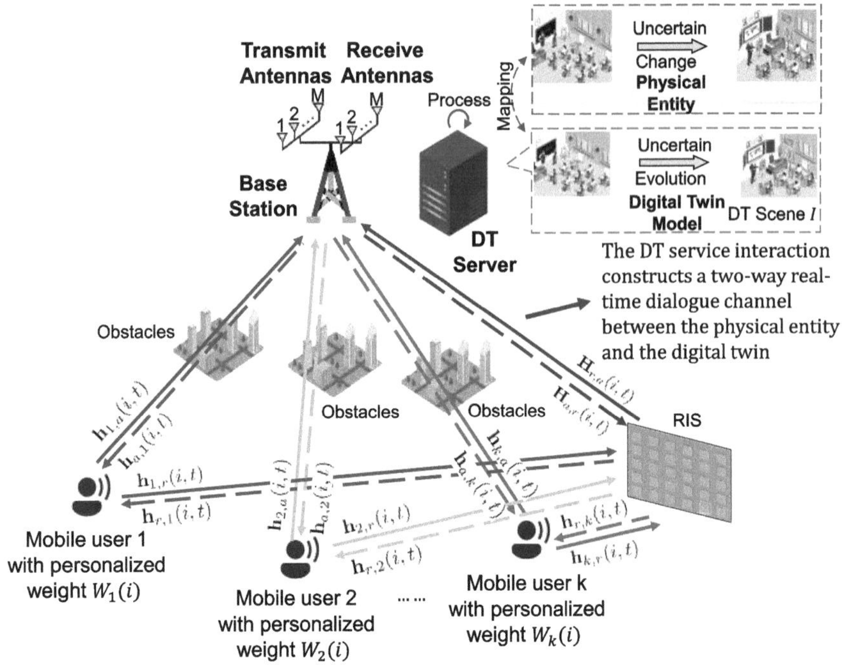

Fig. 4.1 An illustration of the considered RIS-assisted DT interaction system. In this system, mobile users first upload their interaction signals to the DT server through effective uplink channels. Then, the DT server, which hosts the DT model, generates and renders the customized feedback, transmitting it back to the mobile users via effective downlink channels. It is worth noting that, the DT model continuously evolves with uncertainties during the DT interactions

terparts. During such DT interactions, mobile users upload their interaction signals to the DT server, including signals of locations, motions, facial expressions, voices, etc. The DT server generates customized feedback (e.g., text, image, audio, video) based on the received interaction signals, and renders the feedback signals before transmitting them back to mobile users [7]. To ensure seamless connectivities and real-time DT interaction services, an RIS is enabled to assist uplink and downlink transmissions between the DT server and mobile users. By adjusting passive reflecting elements, the RIS can intelligently reflect both uplink and downlink radio signals simultaneously [35].

It is worth noting that the status of the physical entity in the real world is continuously changing, which drives the uncertain evolution of the corresponding DT model hosted by the DT server. This indicates that the DT scene which mobile users engage in, may continuously evolve. To depict such a unique feature, we define a set $\mathcal{I}$ to include all possible DT scenes, and consider that its cardinality is growing with the evolution, i.e., $|\mathcal{I}| \rightarrow +\infty$. Furthermore, we allow mobile users to interact with the DT server under each DT scene $i \in \mathcal{I}$ multiple times by a given frequency $\mathcal{F}_i$. Then, we can divide the period of each DT scene $i \in \mathcal{I}$ into $T_i = 1/\mathcal{F}_i$ time slots, denoted as $\mathcal{T}_i = \{1, 2, ..., T_i\}$, and within each time slot $t \in \mathcal{T}$, a round-trip interaction is conducted between mobile users and the DT server. Ideally, when engaging in any DT scene $i \in \mathcal{I}$, each mobile user desires both high levels of subjective experience (i.e., the feeling of authenticity) and objective experience (i.e., ultra-low latency in communications and computations) [32, 67]. However, for resource-limited wireless systems, there is an inherent trade-off between these two goals. Specifically, optimizing the subjective (objective) experience commonly requires to enhance (lower) feedback signal resolutions, leading to the increase (decrease) of computation and communication burdens, in turn diminishing the objective (subjective) experience. Meanwhile, depending on particular applications, each mobile user $k \in \mathcal{K}$ may have a personalized weight for objective and subjective experiences [32, 39], varying across different DT scenes, defined as $\mathcal{W}_k(i) = (\varpi_k^\iota(i), \varpi_k^\epsilon(i))$, where $\varpi_k^\iota(i)$ and $\varpi_k^\epsilon(i)$ represent the degrees of mobile user k's attentions to objective and subjective experiences towards DT scene $i \in \mathcal{I}$, respectively, and obviously $\varpi_k^\iota(i) + \varpi_k^\epsilon(i) = 1$.

Overall, we aim to jointly optimize uplink and downlink resource allocations in RIS-assisted wireless communications, along with feedback signal resolution and computational configurations, for all mobile users to balance their objective and subjective experiences of DT interactions. Taking into account the uncertain DT evolution and the resulted unbounded growing amount of different scene-specific problems, we target to design and train a generalized approach with a low complexity, which can automatically produce(i) the phase shift at the RIS, (ii) the receive beamforming at the DT server, (iii) the transmit beamforming at the DT server, (iv) the feedback signal rendering resolution configuration, and (v) the computation resource configuration at the DT server, for all mobile users within each time slot $t \in \mathcal{T}_i$ when engaging in every DT scene $i \in \mathcal{I}$ but not a specific one only. For convenience, the main notations used in this paper are listed in Table 4.1.

4.3.2 Uplink/Downlink Transmissions in DT Interaction

Within any time slot $t \in \mathcal{T}_i$ when engaging in DT scene $i \in \mathcal{I}$, each mobile user $k \in \mathcal{K}$, located at $\mathbf{q_k}(i) = [x_k(i), y_k(i), z_k(i)]$, transmits its interaction signals to the DT server through an effective uplink channel $\mathbf{h}_{k,a}^{UL}(i,t) \in \mathbb{C}^{M\times 1}$, while obtaining feedback signals from the DT server through an effective downlink channel $\mathbf{h}_{a,k}^{DL}(i,t) \in \mathbb{C}^{M\times 1}$, both assisted by RIS, which can be respectively expressed as

$$\mathbf{h}_{k,a}^{UL}(i,t) = \underbrace{\mathbf{h}_{k,a}(i,t)}_{\text{Mobile user} \rightarrow \text{DT server}} + \underbrace{\mathbf{H}_{r,a}(i,t)\mathbf{\Theta}(i,t)\mathbf{h}_{k,r}(i,t)}_{\text{Mobile user}\rightarrow\text{RIS}\rightarrow\text{DT server}}, \tag{4.1}$$

$$\mathbf{h}_{a,k}^{DL}(i,t) = \underbrace{\mathbf{h}_{a,k}(i,t)}_{\text{DT server}\rightarrow\text{Mobile user}} + \underbrace{\mathbf{H}_{a,r}(i,t)\mathbf{\Theta}(i,t)\mathbf{h}_{r,k}(i,t)}_{\text{DT server}\rightarrow\text{RIS}\rightarrow\text{Mobile user}}, \tag{4.2}$$

Table 4.1 Important notations in this chapter

Symbol	Meaning
$\mathcal{K}, \mathcal{I}$	Set of mobile users and DT scenes
M	Amount of receive/transmit antennas
N	Amount of RIS r's reflecting elements
$\mathbf{q_a}, \mathbf{q_r}, \mathbf{q_k}(i, t)$	Locations of a and r, and $k \in \mathcal{K}$ under $i \in \mathcal{I}$ at time slot $t \in \mathcal{T}_i$
$\mathcal{T}_i$	Set of time slots under $i \in \mathcal{I}$
$\mathcal{W}_k(i)$	k's personalized weights for subjective and objective experiences under $i \in \mathcal{I}$
$QoE(\mathcal{W}_k(i), t)$	QoE of $k \in \mathcal{K}$ within $t \in \mathcal{T}_i$ under $i \in \mathcal{I}$
$\mathbf{h}_{k,a}^{UL}(i, t), \mathbf{h}_{k,a}^{DL}(i, t)$	Effective uplink and downlink channels for $k \in \mathcal{K}$ within $t \in \mathcal{T}_i$ under $i \in \mathcal{I}$
$\mathbf{\Theta}(i, t)$	Phase shift matrix within $t \in \mathcal{T}_i$ under $i \in \mathcal{I}$
$\mathbf{h}_{k,a}(i, t), \mathbf{h}_{a,k}(i, t)$	Channels between $k \in \mathcal{K}$ and a of $\mathbf{h}_{k,a}^{UL}(i, t)$ and $\mathbf{h}_{k,a}^{DL}(i, t)$, respectively
$\mathbf{h}_{k,r}(i, t), \mathbf{h}_{r,k}(i, t)$	Channels between $k \in \mathcal{K}$ and r of $\mathbf{h}_{k,a}^{UL}(i, t)$ and $\mathbf{h}_{k,a}^{DL}(i, t)$, respectively
$\mathbf{H}_{r,a}(i, t), \mathbf{H}_{a,r}(i, t)$	Channels between r and a of $\mathbf{h}_{k,a}^{UL}(i, t)$ and $\mathbf{h}_{k,a}^{DL}(i, t)$, respectively
$v_k^{UL}(i, t), v_k^{DL}(i, t)$	Transmission rates of uplink and downlink interaction between $k \in \mathcal{K}$ and a within $t \in \mathcal{T}_i$ under $i \in \mathcal{I}$
$\gamma_k^{UL}(i, t), \gamma_k^{DL}(i, t)$	Uplink and downlink SINR when $k \in \mathcal{K}$ engaging in $i \in \mathcal{I}$ within $t \in \mathcal{T}_i$
$D_k^{UL}(i)$	Size of k's interaction signal under $i \in \mathcal{I}$
$E_k(i, t)$	Resolution of feedback signal for $k \in \mathcal{K}$ within $t \in \mathcal{T}_i$ under $i \in \mathcal{I}$
μ	The amount of required computing resource for processing 1-bit feedback signals
$f_k(i, t)$	Computation frequency allocated to $k \in \mathcal{K}$ within $t \in \mathcal{T}_i$ under $i \in \mathcal{I}$
$\mathcal{L}_k^{UL}(i, t), \mathcal{L}_k^{PRO}(i, t), \mathcal{L}_k^{DL}(i, t), \mathcal{L}_k(i, t)$	Uplink latency, processing latency, downlink latency, and interaction round-trip latency of $k \in \mathcal{K}$ within $t \in \mathcal{T}_i$ under $i \in \mathcal{I}$
$\mathcal{L}_{max}$	Maximal tolerable round-trip latency
E_{min}	The minimum resolution that users demand
C	Computational capacity of the DT server
p^{max}	Maximum transmit power of DT server
$\mathbf{V}(i, t), \mathbf{W}(i, t)$	Receive and transmit beamforming matrices at DT server within $t \in \mathcal{T}_i$ under $i \in \mathcal{I}$

where vectors $\mathbf{h}_{k,a}(i, t)$ and $\mathbf{h}_{a,k}(i, t)$ specify the direct uplink and downlink channels between mobile user $k \in \mathcal{K}$ and the DT server, respectively; $\mathbf{H}_{r,a}(i, t)\mathbf{\Theta}(i, t)\mathbf{h}_{k,r}(i, t)$ and $\mathbf{H}_{a,r}(i, t)\mathbf{\Theta}(i, t)\mathbf{h}_{r,k}(i, t)$ represent the indirect uplink and downlink channels via RIS between mobile user $k \in \mathcal{K}$ and the DT server, respectively, where matrices $\mathbf{H}_{r,a}(i,t)$, $\mathbf{H}_{a,r}(i,t)$ indicate the uplink and downlink channels between the RIS and the DT server, respectively, and vectors $\mathbf{h}_{k,r}(i,t)$, $\mathbf{h}_{r,k}(i,t)$ stand for the uplink and downlink channels between mobile user k and the RIS, respectively. $\mathbf{\Theta}(i,t) \triangleq diag(A_1(i,t)e^{j\theta_1(i,t)}, ..., A_N(i,t)e^{j\theta_N(i,t)}) \in \mathbb{C}^{N\times N}$ is the phase shift matrix of the RIS, where $A_n(i,t)$ and $\theta_n(i,t)$ represent the amplitude reflection coefficient and the phase shift induced by nth RIS element. Similar to [11, 25], we consider that ideal reflections can be done by the RIS, which satisfies $|A_n(i,t)e^{j\theta_n(i,t)}|^2 = 1$, i.e., $A_n(i,t) = 1, n \in \mathcal{N}, i \in \mathcal{I}, t \in \mathcal{T}_i$. To realize more precise phase control, the continuous phase shift is enabled, and thus we have $\theta_n(i,t) \in [0, 2\pi), n \in \mathcal{N}$, $i \in \mathcal{I}, t \in \mathcal{T}_i$.

In the following, we elaborate the detailed modeling of $\mathbf{h}^{UL}_{k,a}(i,t)$ and $\mathbf{h}^{DL}_{a,k}(i,t)$, $k \in \mathcal{K}, i \in \mathcal{I}, t \in \mathcal{T}_i$, wherein the corresponding transmission rate are also derived. According to [61, 64], the scattering links between the DT server and mobile user $k \in \mathcal{K}$ are modeled by Rayleigh fading, and other RIS-related links can be seen as scatter-free line-of-sight (LoS), modeled by Rician fading. Furthermore, it is worth mentioning that the wireless communication channels are assumed to be quasi-static within each time slot $t \in \mathcal{T}_i$ under any DT scene $i \in \mathcal{I}$, but can be time-varying over different slots [14, 33, 34].

4.3.2.1 Effective Uplink Channel

For direct uplink communications of the DT interaction within each time slot $t \in \mathcal{T}_i$ under DT scene $i \in \mathcal{I}$, the channel between mobile user $k \in \mathcal{K}$ and the DT server can be modeled as

$$\mathbf{h}_{k,a}(i,t) = \sqrt{\rho d_{k,a}^{-\alpha_{k,a}}(i)}\mathbf{h}^{NLoS}_{k,a}(i,t) \in \mathbb{C}^{M\times 1}, \tag{4.3}$$

where ρ is the path loss coefficient, $\alpha_{k,a}$ is the path loss exponent, $d_{k,a}(i) = \sqrt{||\mathbf{q_k}(i) - \mathbf{q_a}||^2}$ is the distance between mobile user k and the DT server, and $\mathbf{h}^{NLoS}_{k,a}(i,t) \in \mathbb{C}^{M\times 1}$ is the Non-LoS (NLoS) component of this link, with each entry obeying $\mathcal{CN}(0,1)$ [17].

For indirect uplink communications of the DT interaction within each time slot $t \in \mathcal{T}_i$ under DT scene $i \in \mathcal{I}$, the channel between mobile user $k \in \mathcal{K}$ and the RIS can be modeled as

$$\mathbf{h}_{k,r}(i,t) = \sqrt{\rho d_{k,r}^{-\alpha_{k,r}}(i)}\Big(\sqrt{\frac{G_{k,r}}{G_{k,r}+1}}\mathbf{h}^{LoS}_{k,r}(i) + \sqrt{\frac{1}{G_{k,r}+1}}\mathbf{h}^{NLoS}_{k,r}(i,t)\Big) \in \mathbb{C}^{N\times 1}, \tag{4.4}$$

where $\alpha_{k,r}$ is the path loss exponent, $d_{k,r}(i) = \sqrt{||\mathbf{q_k}(i) - \mathbf{q_r}||^2}$ is the distance between mobile user k and the RIS, and $\mathbf{h}^{NLoS}_{k,r}(i,t) \in \mathbb{C}^{N\times 1}$ is the NLoS component of this link, with each entry obeying $\mathcal{CN}(0,1)$, and $G_{k,r}$ is the Rician factor.

Besides, $\mathbf{h}_{k,r}^{LoS}(i) = [1, ..., e^{-j\frac{2\pi d}{\lambda}(N-1)\varphi_{k,r}^{AoA}(i)}]^T \in \mathbb{C}^{N\times 1}$ is the LoS component of this link, where $\varphi_{k,r}^{AoA}(i) = \frac{x_r - x_k(i)}{||\mathbf{q}_k(i) - \mathbf{q}_r||}$ is the cosine of the angle of arrival (AoA) of the signals from mobile user k to the RIS. Besides, d and λ represent element separation and carrier wavelength, respectively [35]. Furthermore, the channel between the RIS and the DT server can be modeled as

$$\mathbf{H}_{r,a}(i,t) = \sqrt{\rho d_{r,a}^{-\alpha_{r,a}}}\left(\sqrt{\frac{G_{r,a}}{G_{r,a}+1}}\mathbf{H}_{r,a}^{LoS} + \sqrt{\frac{1}{G_{r,a}+1}}\mathbf{H}_{r,a}^{NLoS}(i,t)\right) \in \mathbb{C}^{M\times N}, \tag{4.5}$$

where $\alpha_{r,a}$ is the path loss exponent, $d_{r,a} = \sqrt{||\mathbf{q_r} - \mathbf{q_a}||^2}$ is the distance between the RIS and the DT server, $\mathbf{H}_{r,a}^{NLoS}(i,t) \in \mathbb{C}^{M\times N}$ is the NLoS component of this link, with each entry obeying $\mathcal{CN}(0,1)$, and $G_{r,a}$ is the Rician factor. Besides, $\mathbf{H}_{r,a}^{LoS} = \mathbf{f}_M \cdot \mathbf{f}_N^T \in \mathbb{C}^{M\times N}$ is the LoS component, where $\mathbf{f}_M = [1, ..., e^{-j\frac{2\pi d}{\lambda}(M-1)\varphi_{r,a}^{AoA}}]^T \in \mathbb{C}^{M\times 1}$, and $\mathbf{f}_N = [1, ..., e^{-j\frac{2\pi d}{\lambda}(N-1)\varphi_{r,a}^{AoD}}]^T \in \mathbb{C}^{N\times 1}$ denote array responses, in which $\varphi_{r,a}^{AoA} = \frac{x_a - x_r}{||\mathbf{q}_r - \mathbf{q}_a||}$ and $\varphi_{r,a}^{AoD} = \frac{x_a - x_r}{||\mathbf{q}_r - \mathbf{q}_a||}$ are the cosine of the AoA and angle of departure (AoD) of the signal from the RIS to the DT server, respectively [61].

Substituting (4.3)–(4.5) into (4.1), the effective uplink channel $\mathbf{h}_{k,a}^{UL}(i,t)$ between mobile user $k \in \mathcal{K}$ and the DT server within time slot $t \in \mathcal{T}_i$ when engaging in DT scene $i \in \mathcal{I}$ can be obtained. Then, we can express the interaction signals of mobile user $k \in \mathcal{K}$ received by DT server via $\mathbf{h}_{k,a}^{UL}(i,t)$ as

$$\begin{aligned} y_k^{UL}(i,t) = &\underbrace{\sqrt{p_k^{UL}}(\mathbf{v}_k(i,t))^H \mathbf{h}_{k,a}^{UL}(i,t)\delta_k^{UL}(i,t)}_{\text{Intended Signal}} \\ &+ \underbrace{(\mathbf{v}_k(i,t))^H(\sum\nolimits_{m=1,m\neq k}^{\mathcal{K}} \sqrt{p_m}\mathbf{h}_{m,a}^{UL}(i,t)\delta_m^{UL}(i,t))}_{\text{Inter-user Interference}} \\ &+ \underbrace{(\mathbf{v}_k(i,t))^H(\mathbf{n}_k^{UL}(i,t))}_{\text{AWGN}}, \end{aligned} \tag{4.6}$$

where the first term denotes the intended signal received at the DT server, the second term is the multi-user interference at the DT server, and the third term represents the circularly symmetric complex additive white Gaussian noise (AWGN) at the DT server. Besides, p_k^{UL} is the uplink transmission power of mobile user k, $\delta_k^{UL}(i,t) \sim \mathcal{CN}(0,1)$ is the uplink interaction information symbol of mobile user k, $\mathbf{v}_k(i,t) \in \mathbb{C}^{M\times 1}$ is the receive beamforming vector at the DT server for mobile user k, and $\mathbf{n}_k^{UL}(i,t) \in \mathbb{C}^{M\times 1}$, with each entry obeying $\mathcal{CN}(0, (\sigma_k^{UL})^2)$.

Accordingly, the transmission rate of uplink interaction between mobile user $k \in \mathcal{K}$ and the DT server within time slot $t \in \mathcal{T}_i$ when engaging in DT scene i can be calculated as

$$\upsilon_k^{UL}(i,t) = b\log_2(1 + \gamma_k^{UL}(i,t)), \tag{4.7}$$

where b is the uplink channel bandwidth, and $\gamma_k^{UL}(i,t)$ is the uplink signal-to-interference-plus-noise ratio (SINR), which can be further expanded as

$$\gamma_k^{UL}(i,t) = \frac{p_k^{UL}|(\mathbf{h}_{k,a}^{UL}(i,t))^H \mathbf{v}_k(i,t)|^2}{(\mathbf{v}_k(i,t))^H((\sigma_k^{UL})^2 \mathbf{I}_M + \sum_{m=1,m\neq k}^{\mathcal{K}} p_m^{UL}\mathbf{h}_{m,a}^{UL}(i,t)(\mathbf{h}_{m,a}^{UL}(i,t))^H \mathbf{v}_k(i,t)}. \tag{4.8}$$

4.3.2.2 Effective Downlink Channel

For direct downlink communications of the DT interaction within each time slot $t \in \mathcal{T}_i$ under DT scene $i \in \mathcal{I}$, the channel between the DT server and mobile user $k \in \mathcal{K}$ can be modeled as

$$\mathbf{h}_{a,k}(i,t) = \sqrt{\rho d_{a,k}^{-\alpha_{a,k}}(i)}\mathbf{h}_{a,k}^{NLoS}(i,t) \in \mathbb{C}^{M\times 1}, \tag{4.9}$$

where $\alpha_{a,k}$ is the path loss exponent, $d_{a,k}(i) = d_{k,a}(i)$, and $\mathbf{h}_{a,k}^{NLoS}(i,t) \in \mathbb{C}^{M\times 1}$ is the NLoS component of this link, with each entry obeying $\mathcal{CN}(0,1)$.

For indirect downlink communications of DT interactions within each time slot $t \in \mathcal{T}_i$ under scene $i \in \mathcal{I}$, the channel between the DT server and the RIS is modeled as

$$\mathbf{H}_{a,r}(i,t) = \sqrt{\rho d_{a,r}^{-\alpha_{a,r}}}\left(\sqrt{\frac{G_{a,r}}{G_{a,r}+1}}\mathbf{H}_{a,r}^{LoS} + \sqrt{\frac{1}{G_{a,r}+1}}\mathbf{H}_{a,r}^{NLoS}(i,t)\right) \in \mathbb{C}^{M\times N}, \tag{4.10}$$

where $\alpha_{r,a}$ is the path loss exponent, $d_{r,a} = d_{a,r}$, $\mathbf{H}_{a,r}^{NLoS}(i,t) \in \mathbb{C}^{M\times N}$ is the NLoS component of this link, with each entry obeying $\mathcal{CN}(0,1)$, and $G_{a,r}$ is the Rician factor. Besides, $\mathbf{H}_{a,r}^{LoS} = \mathbf{f}_M \cdot \mathbf{f}_N^T \in \mathbb{C}^{M\times N}$ is the LoS component, where $\mathbf{f}_M = [1, ..., e^{-j\frac{2\pi d}{\lambda}(M-1)\varphi_{a,r}^{AoD}}]^T \in \mathbb{C}^{M\times 1}$, and $\mathbf{f}_N = [1, ..., e^{-j\frac{2\pi d}{\lambda}(N-1)\varphi_{a,r}^{AoA}}]^T \in \mathbb{C}^{N\times 1}$ denote array responses, in which $\varphi_{a,r}^{AoD} = \frac{x_r - x_a}{||\mathbf{q}_a - \mathbf{q}_r||}$ and $\varphi_{a,r}^{AoA} = \frac{x_r - x_a}{||\mathbf{q}_a - \mathbf{q}_r||}$ are the cosine of the AoD and AoA of the signal from the DT server to the RIS, respectively [36]. Furthermore, the channel between the RIS and mobile user $k \in \mathcal{K}$ can be modeled as

$$\mathbf{h}_{r,k}(i,t) = \sqrt{\rho d_{r,k}^{-\alpha_{r,k}}(i)}\left(\sqrt{\frac{G_{r,k}}{G_{r,k}+1}}\mathbf{h}_{r,k}^{LoS}(i) + \sqrt{\frac{1}{G_{r,k}+1}}\mathbf{h}_{r,k}^{NLoS}(i,t)\right) \in \mathbb{C}^{N\times 1}, \tag{4.11}$$

where $\alpha_{r,k}$ is the path loss exponent, $d_{r,k}(i) = d_{k,r}(i)$, $\mathbf{h}_{r,k}^{NLoS}(i,t) \in \mathbb{C}^{N\times 1}$ is the NLoS component of this link, with each entry obeying $\mathcal{CN}(0,1)$, and $G_{r,k}$ is the Rician factor. Meanwhile, $\mathbf{h}_{r,k}^{LoS}(i) = [1, ..., e^{-j\frac{2\pi d}{\lambda}(N-1)\varphi_{r,k}^{AoD}(i)}]^T \in \mathbb{C}^{N\times 1}$ is the LoS component of this link, where $\varphi_{r,k}^{AoD}(i) = \frac{x_k(i) - x_r}{||\mathbf{q}_r - \mathbf{q}_k(i)||}$ is the cosine of the AoD of the signals from the RIS to mobile user $k \in \mathcal{K}$.

Substituting (4.9)–(4.11) into (4.2), the effective downlink channel $\mathbf{h}_{a,k}^{DL}(i,t)$ between DT server and mobile user $k \in \mathcal{K}$ within time slot $t \in \mathcal{T}_i$ when engaging in DT scene $i \in \mathcal{I}$ can be obtained. Then, we can express the downlink feedback signals received by mobile user $k \in \mathcal{K}$ as

$$\begin{aligned} y_k^{DL}(i,t) = & \underbrace{(\mathbf{h}_{a,k}^{DL}(i,t))^H \mathbf{w}_k(i,t)\delta_k^{DL}(i,t)}_{\text{Intended Signal}} \\ & + \underbrace{\sum\nolimits_{m=1,m\neq k}^{\mathcal{K}} (\mathbf{h}_{a,k}^{DL}(i,t))^H \mathbf{w}_m(i,t)\delta_m^{DL}(i,t)}_{\text{Inter-user Interference}} + \underbrace{n_k^{DL}(i,t)}_{\text{AWGN}}, \end{aligned} \tag{4.12}$$

where $\mathbf{w}_k(i,t) \in \mathbb{C}^{M\times 1}$, $\delta_k^{DL}(i,t) \sim \mathcal{CN}(0,1)$, and $n_k^{DL}(i,t) \sim \mathcal{CN}(0,(\sigma_k^{DL})^2)$ are respectively the downlink transmit beamforming vector at the DT server, the downlink feedback information symbol at mobile user k, and AWGN at mobile user k.

Accordingly, the transmission rate of downlink interaction between the DT server and mobile user $k \in \mathcal{K}$ within time slot $t \in \mathcal{T}_i$ when engaging in DT scene $i \in \mathcal{I}$ can be calculated as

$$v_k^{DL}(i,t) = b\log_2(1+\gamma_k^{DL}(i,t)), \tag{4.13}$$

where $\gamma_k^{DL}(i,t)$ is the downlink SINR, which can be further expanded as

$$\gamma_k^{DL}(i,t) = \frac{|(\mathbf{h}_{a,k}^{DL}(i,t))^H \mathbf{w}_k(i,t)|^2}{(\sigma_k^{DL})^2 + \sum\nolimits_{m=1,m\neq k}^{\mathcal{K}} |(\mathbf{h}_{a,k}^{DL}(i,t))^H \mathbf{w}_m(i,t)|^2}. \tag{4.14}$$

4.3.3 Interaction Round-Trip Latency

Within each time slot $t \in \mathcal{T}_i$ when engaging in any DT scene $i \in \mathcal{I}$, the round-trip latency $\mathcal{L}_k(i,t)$ of the DT interaction between each mobile user $k \in \mathcal{K}$ and the DT server consists of uplink latency, rendering latency and downlink latency. Specifically, each mobile user $k \in \mathcal{K}$ first transmits its interaction signal with size $D_k^{UL}(i)$ via the effective uplink channel $\mathbf{h}_{k,a}^{UL}(i,t)$ to the DT server, resulting in the uplink latency calculated as

$$\mathcal{L}_k^{UL}(i,t) = \frac{D_k^{UL}(i)}{v_k^{UL}(i,t)}. \tag{4.15}$$

Then, the DT server generates and renders the feedback signal for mobile user k with the resolution $E_k(i,t)$, which is optimized according to its personalized weight $\mathcal{W}_k(i)$. The processing (including feedback signal generation and rendering) latency for this can be calculated as

$$\mathcal{L}_k^{PRO}(i,t) = \frac{\xi\mu E_k(i,t)}{f_k(i,t)}, \tag{4.16}$$

where ξ is the per-sample data size in bits, $E_k(i,t)$ is in samples (e.g., pixels of the generated images), μ is the amount of required computing resource (i.e., the number of CPU cycles) for processing 1-bit feedback signals, and $f_k(i,t)$ is the allocated computation frequency.

Subsequently, the rendered feedback signal with data size $\varsigma E_k(i,t)$ should be transmitted back to mobile user $k \in \mathcal{K}$ via the effective downlink channel $\mathbf{h}_{k,a}^{DL}(i,t)$, where ς is the per-sample data size of the processed feedback signal, leading to the downlink latency calculated as

$$\mathcal{L}_k^{DL}(i,t) = \frac{\varsigma E_k(i,t)}{\upsilon_k^{DL}(i,t)}. \tag{4.17}$$

To sum up, the interaction round-trip latency of mobile user $k \in \mathcal{K}$ within each time slot $t \in \mathcal{T}_i$ when engaging in DT scene $i \in \mathcal{I}$ becomes

$$\mathcal{L}_k(i,t) = \mathcal{L}_k^{UL}(i,t) + \mathcal{L}_k^{PRO}(i,t) + \mathcal{L}_k^{DL}(i,t). \tag{4.18}$$

4.3.4 QoE Model Analysis for DT Interaction

To jointly evaluate the quality of each mobile user's objective and subjective experiences of DT interaction within each time slot $t \in \mathcal{T}_i$ under any DT scene $i \in \mathcal{I}$, we introduce a novel user-centric QoE metric, i.e., $QoE(\mathcal{W}_k(i), t), \forall k \in \mathcal{K}$. Particularly, we consider two key performance indicators, i.e., the Weber-Fechner Law based human perception quality $\mathcal{E}_k(i,t)$ and the interaction round-trip latency $\mathcal{L}_k(i,t)$, to quantify subjective and objective experiences of each mobile user $k \in \mathcal{K}$, respectively. By combining $\mathcal{E}_k(i,t)$ and $\mathcal{L}_k(i,t)$, which are bridged by personalized weight $\mathcal{W}_k(i)$, each mobile user k's QoE of the DT interaction can be obtained.

Originally, the Weber-Fechner Law characterizes the relationship between the human perception and the relative change of the physical stimulus intensity [50]. Analogously, for the DT interaction services, we can model the physical stimulus intensity as the rendering resolution $E_k(i,t)$, which determines the authenticity of the service. According to the Weber-Fechner Law, the differential human perception is directly proportional to the relative change in stimulus, and thereby, we can express the differential subjective experience $d\mathcal{E}_k(i,t)$ of each mobile user $k \in \mathcal{K}$ within each time slot $t \in \mathcal{T}_i$ under DT scene $i \in \mathcal{I}$ as

$$d\mathcal{E}_k(i,t) = \frac{dE_k(i,t)}{E_k(i,t)}. \tag{4.19}$$

Then, $\mathcal{E}_k(i,t)$ of each mobile user $k \in \mathcal{K}$ can be derived as

$$\mathcal{E}_k(i,t) = \ln\left(\frac{E_k(i,t)}{E_{min}}\right), \tag{4.20}$$

where E_{min} is minimum resolution that users demand.

By integrating objective and subjective experience metrics through the personalized weight $\mathcal{W}_k(i)$, we define $QoE(\mathcal{W}_k(i), t)$ for each mobile user $k \in \mathcal{K}$ within each time slot $t \in \mathcal{T}_i$ under DT scene $i \in \mathcal{I}$ as

$$QoE(\mathcal{W}_k(i), t) = \varpi_k^{\iota}(i)F^{'}(\mathcal{L}_k(i,t)) + \varpi_k^{\epsilon}(i)F^{''}(\mathcal{E}_k(i,t)), \tag{4.21}$$

where

$$F^{'}(\mathcal{L}_k(i,t)) = 1 - \frac{\mathcal{L}_k(i,t)}{\mathcal{L}_{max}},\ F^{''}(\mathcal{E}_k(i,t)) = \frac{\mathcal{E}_k(i,t)}{\mathcal{E}_{max}}. \tag{4.22}$$

Here, functions $F^{'}(\cdot)$ and $F^{''}(\cdot)$ are used to characterize mobile users' satisfaction degrees in terms of objective and subjective experiences, respectively. $\mathcal{L}_{max}$ is the maximal interaction round-trip latency that each mobile user can tolerate, and $\mathcal{E}_{max}$ is the maximal perception quality that each mobile user can achieve, both of which can be obtained from empirical studies [10, 40].

4.3.5 Problem Formulation

For engaging in every DT scene $i \in \mathcal{I}$, we aim to maximize the sum of all mobile users' QoE of DT interaction across all time slots, i.e., $\sum_{t\in\mathcal{T}_i}\sum_{k\in\mathcal{K}} QoE(\mathcal{W}_k(i), t)$, by jointly optimizing the phase shift matrix $\mathbf{\Theta}(i,t)$ at the RIS, receive beamforming matrix $\mathbf{V}(i,t) := \{\mathbf{v}_1(i,t), ..., \mathbf{v}_K(i,t)\}$ and transmit beamforming matrix $\mathbf{W}(i,t) := \{\mathbf{w}_1(i,t), ..., \mathbf{w}_K(i,t)\}$ of the DT server, rendering resolution configuration $E_k(i,t)$ and computing resource allocation $f_k(i,t)$ for each mobile user $k \in \mathcal{K}$ within each time slot $t \in \mathcal{T}_i$, i.e.,

$$\mathcal{P}(i): \max_{\mathbf{\Theta}(i,t),\mathbf{V}(i,t),\mathbf{W}(i,t),E_k(i,t),f_k(i,t)} \sum_{t\in\mathcal{T}_i}\sum_{k\in\mathcal{K}} QoE(\mathcal{W}_k(i), t) \tag{4.23}$$

$$s.t.,\ \theta_n(i,t) \in [0, 2\pi), \forall n \in \mathcal{N}, \tag{4.23a}$$

$$E_k(i,t) \in [E_{min}, E_{max}], \forall k \in \mathcal{K}, \tag{4.23b}$$

$$\sum_{k\in\mathcal{K}} f_k(i,t) \le C,\ f_k(i,t) > 0, \tag{4.23c}$$

$$\sum_{k\in\mathcal{K}} ||\mathbf{w}_k(i,t)||^2 \le P^{max}, \tag{4.23d}$$

$$\mathcal{L}_k(i,t) \le \mathcal{L}_{max}, \forall k \in \mathcal{K}, \tag{4.23e}$$

where constraint (4.23a) signifies the continuous phase shift control of RIS elements; constraint (4.23b) indicates the configuration range of the feedback signal resolution for each mobile user; constraint (4.23c) is the computational resource allocation constraint, where C denotes the computational capacity of the DT server; constraint (4.23d) is the downlink transmission power constraint, and P^{max} denotes the maximum transmit power of the DT server; constraint (4.23e) guarantees that the interaction round-trip latency of each mobile user does not exceed a pre-determined threshold.

Since the DT scene that mobile users engage in is continually evolving, the uncertain DT evolution leads to unbounded growing amount of different scene-specific problems, which can be expressed in a series of problems as

$$\mathcal{P} = \{\mathcal{P}(1), \mathcal{P}(2), ..., \mathcal{P}(|\mathcal{I}|)\}_{|\mathcal{I}| \to +\infty}, \tag{4.24}$$

where each $\mathcal{P}(i) \in \mathcal{P}$ is mainly distinguished by mobile users' personalized weights on objective and subjective experiences under different scenes. Obviously, for providing mobile users with high-quality DT interaction services, we should efficiently solve $\mathcal{P}$ for every DT scene instead of $\mathcal{P}(i)$ for one particular DT scene i only. However, solving problem $\mathcal{P}$ is very challenging due to the following reasons. First, even for each individual scene-specific problem $\mathcal{P}(i), \forall i \in \mathcal{I}$, both the objective function (4.23) and constraint (4.23e) involve non-convex quadratic terms, and thus it is NP-hard [1, 42]. Moreover, $\mathcal{P}(i), \forall i \in \mathcal{I}$ is indeed a long-term problem over multiple time slots, requiring an online algorithm to run dynamically. In addition, $\mathcal{P}$ actually consists of a series of problems $\mathcal{P}(i), \forall i \in \mathcal{I}$ in accordance with the uncertain DT evolution, requiring us to constantly re-solve each one of them whenever the engaged scene changes. This further exacerbates the computational complexity of addressing $\mathcal{P}$, making it extremely hard if not impossible to produce solutions for guaranteeing fast-responsive DT interaction services in real-time.

We observe that all scene-specific problems in $\mathcal{P}$ are ordered in sequence and are not completely different but share similar fundamental structures in the problem formulation. These motivate us to potentially utilize the experienced knowledge in solving the historical problems to efficiently solve the new yet similar problems in $\mathcal{P}$ by incorporating the introduced scene-specific information. Decision transformer [9, 20, 65], with the superior capability of tackling dynamic and non-convex optimization problems, may be well-suited for solving each individual $\mathcal{P}(i), \forall i \in \mathcal{I}$. Moreover, inspired by the core idea of prompt-based learning in the recently popular GAI models, such as ChatGPT [2] and DALL-E [49], we may further extend the traditional decision transformer to a prompt-guided decision transformer with the strong generalization in problem-solving, which can be accommodated to new yet similar problems after training on a large set of historical ones. By this way, the scene-specific information of every $\mathcal{P}(i) \in \mathcal{P}$ can be elaborately captured in prompts, enabling the long-term trained prompt-guided decision transformer to quickly and automatically generate corresponding solutions as answers without re-training.

4.4 Prompt-Guided Decision Transformer Integrated with Zero-Forcing Optimization (PG-ZFO)

In this section, we propose a novel GAI-aided approach, namely prompt-guided decision transformer integrated with ZF optimization (PG-ZFO), to efficiently address the formulated problem $\mathcal{P}$, so as to maximize the total QoE of all mobile users in RIS-assisted DT interactions under the uncertain DT evolution.

4.4.1 Solution Overview

We first introduce a problem reformulation, to transform each individual scene-specific problem $\mathcal{P}(i) \in \mathcal{P}$ into an MDP $\mathcal{M}(i) \in \mathcal{M}$, where $\mathcal{M}$ denotes the set of reformulated problems corresponding to $\mathcal{P}$, i.e., $\mathcal{M} = \{\mathcal{M}(1), \mathcal{M}(2), \ldots, \mathcal{M}(|\mathcal{I}|)\}_{|\mathcal{I}| \to +\infty}$, making it suitable as the input of the decision transformer, which is also the foundation of the proposed PG-ZFO. Then, to efficiently solve each MDP $\mathcal{M}(i) \in \mathcal{M}$, we particularly design a prompt, which implicitly captures the reward function and the transition probability function of MDP $\mathcal{M}(i) \in \mathcal{M}$, conveying the scene-specific information of $\mathcal{P}(i) \in \mathcal{P}$. Based on this, we build a prompt-guided decision transformer to generate phase shift matrix, rendering resolution configuration and computing resource allocation, instructed by the designed prompt, given the current state of MDP $\mathcal{M}(i) \in \mathcal{M}$. On top of this, we further develop a ZF-based optimization algorithm to derive the high-dimensional receive/transmit beamforming matrix, by taking the output of the prompt-guided decision transformer as the input. The decisions made by prompt-guided decision transformer and ZF-based optimization algorithm form an action, i.e., the solution in the current state, which is applied back to $\mathcal{M}(i) \in \mathcal{M}$ to attain the reward and next state. Such decision-making process is iteratively conducted for $\mathcal{M}(i) \in \mathcal{M}$ over all time slots within $\mathcal{T}_i, \forall i \in \mathcal{I}$, and thus $\mathcal{P}(i) \in \mathcal{P}$ is solved.

Unlike the conventional approaches for solving MDPs [38, 55], the proposed PG-ZFO does not need to solve each MDP $\mathcal{M}(i) \in \mathcal{M}$ from the scratch with re-training. Instead, PG-ZFO is trained on a dataset including decision-making processes over diverse historical scene-specific problems, with the target of learning a generalized solution policy for handling any unseen scene-specific problems. During its offline training, the prompt is carefully extracted from the dataset, serving as a prefix to the input and augmenting the learning of generalization ability. In its online execution, the well-trained prompt-guided decision transformer is integrated with the developed ZF-based optimization algorithm, running alternatively to solve any new scene-specific problem, with the corresponding prompt automatically obtained in an online manner. An overview of PG-ZFO is illustrated in Fig. 4.2.

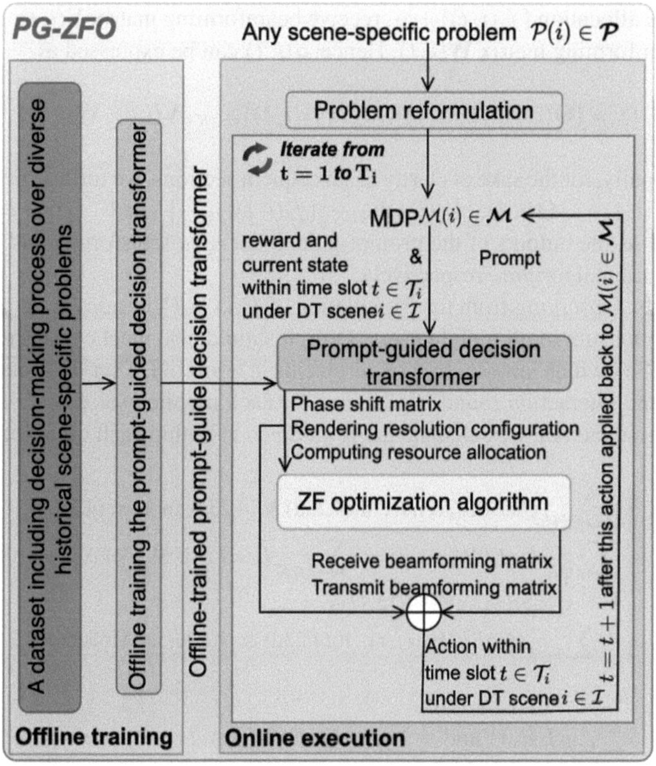

Fig. 4.2 The overview of the proposed PG-ZFO

4.4.2 Problem Reformulation

Here, we reformulate each scene-specific problem $\mathcal{P}(i) \in \boldsymbol{\mathcal{P}}$ into an MDP $\mathcal{M}(i) \in \boldsymbol{\mathcal{M}}$ over $\mathcal{T}_i$, $\forall i \in \mathcal{I}$. Each MDP $\mathcal{M}(i) \in \boldsymbol{\mathcal{M}}$ consists of a state space, an action space, a reward function and a transition probability function, which are described as follows.

- **States**: Within each time slot $t \in \mathcal{T}_i$, the current state $s(i,t)$ includes all mobile users' personalized weights $\{\mathcal{W}_k(i)\}_{\forall k \in \mathcal{K}}$, data size of interaction signals $\{D_k^{UL}(i)\}_{\forall k \in \mathcal{K}}$, effective uplink and downlink channels $\{\mathbf{h}_{k,a}^{UL}(i,t), \mathbf{h}_{a,k}^{DL}(i,t)\}_{\forall k \in \mathcal{K}}$, and QoE $\{QoE\ (\mathcal{W}_k(i), t-1)\}_{\forall k \in \mathcal{K}}$ achieved in the previous time slot $t-1$. Then, $s(i,t)$ can be represented as

$$s(i,t) = \{\mathcal{W}_k(i), D_k^{UL}(i), \mathbf{h}_{k,a}^{UL}(i,t), \mathbf{h}_{a,k}^{DL}(i,t), QoE(\mathcal{W}_k(i), t-1)\}_{\forall k \in \mathcal{K}}. \tag{4.25}$$

- **Actions**: Within each time slot $t \in \mathcal{T}_i$, the action $a(i,t)$ consists of phase shift matrix $\boldsymbol{\Theta}(i,t)$, rendering resolution configuration $\{E_k(i,t)\}_{\forall k \in \mathcal{K}}$, computing resource allocation $\{f_k(i,t)\}_{\forall k \in \mathcal{K}}$, receive beamforming matrix $\mathbf{V}(i,t)$, and transmit beamforming matrix $\mathbf{W}(i,t)$. Hence, $a(i,t)$ can be expressed as

$$a(i,t) = \{\boldsymbol{\Theta}(i,t), \{E_k(i,t)\}_{\forall k \in \mathcal{K}}, \{f_k(i,t)\}_{\forall k \in \mathcal{K}}, \mathbf{V}(i,t), \mathbf{W}(i,t)\}. \tag{4.26}$$

Additionally, for the sake of clarity in subsequent sections, we further divide $a(i,t)$ into $\vartheta(i,t) = \{\boldsymbol{\Theta}(i,t), \{E_k(i,t)\}_{\forall k \in \mathcal{K}}, \{f_k(i,t)\}_{\forall k \in \mathcal{K}}\}$ and $\kappa(i,t) = \{\mathbf{V}(i,t), \mathbf{W}(i,t)\}$ as the outputs of the prompt-guided decision transformer and ZF-based optimization algorithm, respectively.

- **Rewards**: Observing from the formulation of $\mathcal{P}(i) \in \boldsymbol{\mathcal{P}}$, besides the goal of maximizing the sum of all mobile users' QoE, it should also meet constraints (4.23d) and (4.23e), which impose strict requirements in terms of DT server's total transmit power and interaction round-trip latency for each mobile user $k \in \mathcal{K}$. To characterize these factors, we calculate the reward $r(i,t)$ within each time slot $t \in \mathcal{T}_i$ as

$$r(i,t) = \begin{cases} \sum_{k \in \mathcal{K}} QoE(\mathcal{W}_k(i),t), \text{ if (4.23d) \& (4.23e) are met, (4.27a)} \\ \sum_{k \in \mathcal{K}} QoE(\mathcal{W}_k(i),t) - \delta \sum_{k \in \mathcal{K}} l_k(i,t), \text{ if (4.23e) is violated, (4.27b)} \\ 0, \text{ if (4.23d) is violated, (4.27c)} \\ -\sum_{k \in \mathcal{K}} QoE(\mathcal{W}_k(i),t), \text{ if (4.23d) \& (4.23e) are violated, (4.27d)} \end{cases} \tag{27}$$

where $-\delta \sum_{k \in \mathcal{K}} l_k(i,t)$ and $-\sum_{k \in \mathcal{K}} QoE(\mathcal{W}_k(i),t)$ are penalties for violating constraints (4.23d) and (4.23e), respectively, δ is the penalty coefficient, and the penalty term $l_k(i,t) = 0$ if $\mathcal{L}_k(i) \leq \mathcal{L}_{max}$, while $l_k(i,t) = \mathcal{L}_{max}$ otherwise.
- **State transition probabilities**: The state transition probability from state $s(i,t)$ to another state $s(i,t+1)$ by taking action $a(i,t)$ can be defined as $Pr(i, s(i,t+1)|s(i,t), a(i,t)) \in [0,1]$.

4.4.3 *Prompt-Guided Decision Transformer in PG-ZFO*

4.4.3.1 Design of Prompt

Unlike descriptive text-based prompts commonly adopted in popular GAI models [41, 53], in order to be applicable for decision-making optimizations, we particularly consider to use "decision-making trajectory" in designing the prompt. Such a decision-making trajectory-based prompt can capture the scene-specific information of $\mathcal{P}(i) \in \boldsymbol{\mathcal{P}}$ by reflecting the complex dynamics of each MDP $\mathcal{M}(i) \in \boldsymbol{\mathcal{M}}$.

Specifically, we define that the prompt consists of a sequence of decision-making trajectory tuples, where each tuple contains reward $r_\star(i,t)$, state $s_\star(i,t)$ and decision $\vartheta_\star(i,t)$ stored from solving the decision-making process within any slot $t \in \mathcal{T}_i$ of $\mathcal{M}(i) \in \boldsymbol{\mathcal{M}}$, represented as

$$\tau_\star(i) = (r_\star(i,t), s_\star(i,t), \vartheta_\star(i,t), ..., r_\star(i,t+T_\star), s_\star(i,t+T_\star), \vartheta_\star(i,t+T_\star)), \tag{4.28}$$

where $T_\star < T_i$ stands for the amount of decision-making trajectory tuples stored in the prompt, and subscript $\cdot_\star$ indicates that the element is associated with the prompt. Prompt $\tau_\star(i)$ can express the scene-specific information of $\mathcal{P}(i) \in \boldsymbol{\mathcal{P}}$ by implicitly capturing the reward function and the state transition probability function of $\mathcal{M}(i) \in \boldsymbol{\mathcal{M}}$. Furthermore, to offer the prompt-guided decision transformer a more forward-looking scene-specific information and make the designed prompt suitable for the decision transformer, instead of using reward $r_\star(i,t)$ directly, we adopt the returns-to-go (RTG) to the prompt. RTG represents the cumulative rewards remaining to be obtained from the current time slot t to T_i, i.e., $\hat{r}_\star(i,t) = \hat{R}(i) - \sum_{j=1}^{j=t} r_\star(i,j)$, where $\hat{R}(i) = \sum_{t\in\mathcal{T}_i}\sum_{k\in\mathcal{K}} \varpi_k^{\epsilon}(i) + \varpi_k^{\iota}(i)$ represents the expected maximum total rewards over all time slots in MDP $\mathcal{M}(i)$. Then, by replacing $r_\star(i,t)$ with $\hat{r}_\star(i,t)$, the prompt designed in (4.28) can be reformulated as

$$\tau_\star(i) = (\hat{r}_\star(i,t), s_\star(i,t), \vartheta_\star(i,t), ..., \hat{r}_\star(i,t+T_\star), s_\star(i,t+T_\star), \vartheta_\star(i,t+T_\star)). \tag{4.29}$$

4.4.3.2 Structure and Workflow of Prompt-Guided Decision Transformer

Next, we build the prompt-guided decision transformer based on the conventional decision transformer [9], while adding a set of trainable linear layers to extract the scene-specific information captured within the prompt for instructing the generation of scene-specific decisions. The whole structure of such a prompt-guided decision transformer is shown in Fig. 4.3, and its workflow is explained in the following.

Compute embeddings for input tokens: We combine the prompt $\tau_\star(i)$ of MDP $\mathcal{M}(i) \in \boldsymbol{\mathcal{M}}$ with the most recent decision-making trajectory with a sequence length of L preceding the current time slot $t \in \mathcal{T}_i$ of MDP $\mathcal{M}(i) \in \boldsymbol{\mathcal{M}}$, i.e., $\tau(i,t) = (\hat{r}(i,t-L+1), s(i,t-L+1), \vartheta(i,t-L+1), \ldots, \hat{r}(i,t), s(i,t), \vartheta(i,t))$, to form the input token $In(\tau_\star(i), \tau(i,t))$ for the prompt-guided decision transformer. To map $In(\tau_\star(i), \tau(i,t))$ into a continuous vector space suitable for prompt-guide decision transformer processing, we apply three modality-specific trainable linear layers to respectively embed the RTG, states, and decisions in $\tau_\star(i)$ of $In(\tau_\star(i), \tau(i,t))$, converting them into token embeddings. Additionally, to enable prompt-guided decision transformer to understand the temporal relationships between different time slots in the decision-making process, we utilize a trainable linear layer to add the same positional embeddings on top of token embeddings of RTG, state, and decision within each decision-making trajectory tuple in $\tau_\star(i)$ of $In(\tau_\star(i), \tau(i,t))$. Similarly, we use another three trainable linear layers to embed $\tau(i,t)$ of $In(\tau_\star(i), \tau(i,t))$ into token embeddings, and a trainable linear layer to add the positional embeddings.

Obtain hidden states using the causal transformer: We feed the embedded input token to a causal transformer to obtain the hidden states. Specifically, we utilize X

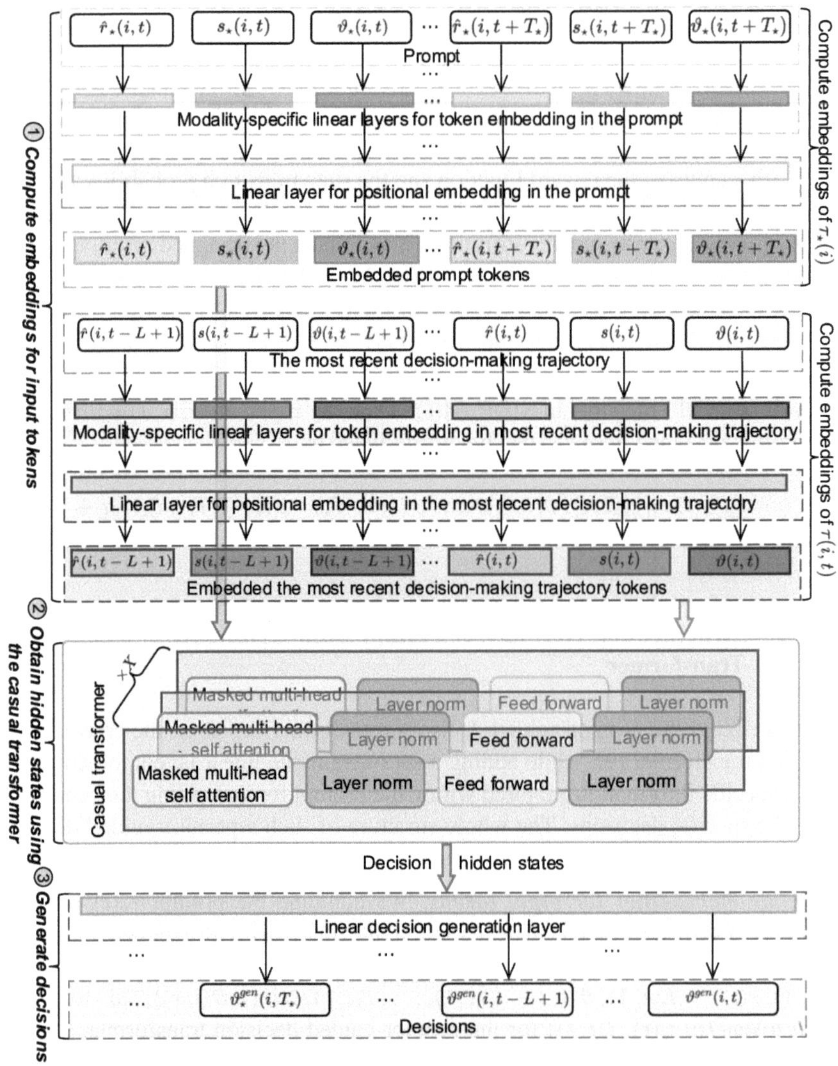

Fig. 4.3 The structure and workflow of prompt-guided decision transformer

stacked identical decoders to build the casual transformer. To effectively represent the scene-specific information encoded in the prompt and the influence of previous decision-making processes on the current decision, in each decoder, we first use a masked multi-head self-attention module to process the embedded input tokens. Subsequently, to enhance the representations generated by the self-attention module, we employ a feed forward layer with position-wise transformations to its outputs. Besides, to stabilize training and facilitate smoother gradient updates during offline training, we apply multiple layer norm [26] throughout each decoder. Moreover, to improve the processing efficiency, we enable the causal transformer to process the embedded input tokens in parallel within each decoder, and sequentially refine them across X stacked decoders. Finally, we obtain the hidden states outputted by the casual transformer.

Generate the decision: To generate the decision $\vartheta^{gen}(i, t)$ under the current state $s(i, t)$ within time slot $t \in \mathcal{T}_i$ for MDP $\mathcal{M} \in \boldsymbol{\mathcal{M}}$, where the superscript $\cdot^{gen}$ indicates that the element is associated with the generated decision, we first extract the decision hidden states from the casual transformer's output, and then apply a trainable linear decision generation layer to process it.

4.4.4 ZF-Based Optimization Algorithm in PG-ZFO

The prompt-guided decision transformer may struggle to generate decisions of the receive beamforming matrix $\mathbf{V}(i, t)$ and transmit beamforming matrix $\boldsymbol{W}(i, t)$ within each time slot $t \in \mathcal{T}_i$ of MDP $\mathcal{M}(i) \in \boldsymbol{\mathcal{M}}$, due to their complex nature and high dimensionality (i.e., $M \times N$ complex space), which could lead to the curse of dimensionality. To address this issue, we further introduce a ZF-based optimization algorithm integrated in the proposed PG-ZFO, which can efficiently determine $\mathbf{V}(i, t)$ and $\boldsymbol{W}(i, t)$ for $t \in \mathcal{T}_i$ of MDP $\mathcal{M}(i) \in \boldsymbol{\mathcal{M}}$, given decision $\vartheta^{gen}(i, t) = \{\boldsymbol{\Theta}(i, t), \{E_k(i, t)\}_{\forall k \in \mathcal{K}}, \{f_k(i, t)\}_{\forall k \in \mathcal{K}}\}$ generated by the prompt-guided decision transformer.

Specifically, within each time slot $t \in \mathcal{T}_i$ of MDP $\mathcal{M}(i) \in \boldsymbol{\mathcal{M}}$, substituting $\vartheta^{gen}(i, t)$ to the original formulation of $\mathcal{P}(i)$, the problem of optimizing decision $\kappa(i, t) = \{\mathbf{V}(i, t), \mathbf{W}(i, t)\}$ for $t \in \mathcal{T}_i$ of MDP $\mathcal{M}(i) \in \boldsymbol{\mathcal{M}}$ can be formulated as

$$\min_{\mathbf{V}(i,t), \mathbf{W}(i,t)} \sum_{k \in \mathcal{K}} \varpi_k^{\iota}(i)(\mathcal{L}_k^{UL}(i, t) + \mathcal{L}_k^{DL}(i, t)) \tag{4.30}$$

$$s.t., \ (4.23d),$$

$$\mathcal{L}_k(i, t) \leq \mathcal{L}_{max} - \mathcal{L}_k^{PRO}(i, t), \forall k \in \mathcal{K}. \tag{4.30a}$$

Note that in the above problem, the beamforming design is extremely difficult because of the inter-user interference in SINR. Hence, for the sake of solution efficiency, similar to [66, 68], we add two constraints for eliminating inter-user interference in beamforming settings, so that problem (4.30) can be reformulated as

$$\min_{\mathbf{V}(i,t), \mathbf{W}(i,t)} \sum_{k \in \mathcal{K}} \varpi_k^{\iota}(i)(\mathcal{L}_k^{UL}(i, t) + \mathcal{L}_k^{DL}(i, t)) \tag{4.31}$$

$$s.t., \ (4.23d), (4.30a),$$

$$|(\mathbf{h}_{m,a}^{UL}(i,t))^H \mathbf{v}_k(i,t)| = 0, \forall k \in \mathcal{K}, m \in \mathcal{K}, m \neq k, \tag{4.31a}$$

$$|(\mathbf{h}_{a,k}^{DL}(i,t))^H \mathbf{w}_m(i,t)| = 0, \forall k \in \mathcal{K}, m \in \mathcal{K}, m \neq k. \tag{4.31b}$$

By analyzing problem (4.31), we can observe that the optimizations of $\mathbf{V}(i,t)$ and $\mathbf{W}(i,t)$ are tightly coupled by constraint (4.30a). For simplicity, we account only for constraint (4.30a) when optimizing $\mathbf{W}(i,t)$, which allow us to decompose problem (4.31) into two subproblems focused on $\mathbf{V}(i,t)$ and $\mathbf{W}(i,t)$, respectively.

Receive beamforming matrix $\mathbf{V}(i,t)$: Accordingly, we decompose problem (4.31) into optimizing $\mathbf{V}(i,t)$ as

$$\min_{\mathbf{V}(i,t)} \sum_{k \in \mathcal{K}} \varpi_k^{\iota}(i) \mathcal{L}_k^{UL}(i,t), \tag{4.32}$$
$$s.t., \ (4.31a).$$

Subproblem (4.32) is still NP-hard because the target involves non-convex quadratic terms. To this end, we employ an efficient algorithm, i.e., the classic ZF algorithm [46, 69], which aims to eliminate inter-user interference by designing receive beamforming vectors $\mathbf{v}_k(i,t), k \in \mathcal{K}$ that are orthogonal to the effective uplink channel of unintended mobile users $k \in \mathcal{K}$, while achieving low uplink latency $\mathcal{L}_k^{UL}(i,t), k \in \mathcal{K}$. Following the detailed steps of the ZF algorithm, we can derive $\mathbf{V}(i,t)$ as

$$\mathbf{V}(i,t) = \mathcal{H}_{UL}(i,t)((\mathcal{H}_{UL}(i,t))^H \mathcal{H}_{UL}(i,t))^{-1}, \tag{4.33}$$

where $\mathcal{H}_{UL}(i,t) = \{\mathbf{h}_{1,a}^{UL}(i,t), ..., \mathbf{h}_{K,a}^{UL}(i,t)\} \in \mathbb{C}^{M \times K}$ represents the matrix of effective uplink channels for all mobile users within time slot $t \in \mathcal{T}_i$ under DT scene $i \in \mathcal{I}$.

Transmit beamforming matrix $\mathbf{W}(i,t)$: Given the receive beamforming matrix $\mathbf{V}(i,t)$ in (4.33), we are able to extract the receive beamforming vector $\mathbf{v}_k(i,t)$ of each mobile user $k \in \mathcal{K}$ within time slot $t \in \mathcal{T}_i$ under DT scene $i \in \mathcal{I}$ from $\mathbf{V}(i,t)$. First, by substituting $\mathbf{v}_k(i,t)$ into (4.15), we can obtain the uplink latency $\mathcal{L}_k^{UL}(i,t)$ of mobile user $k \in \mathcal{K}$ within time slot $t \in \mathcal{T}_i$ under DT scene $i \in \mathcal{I}$ as

$$\mathcal{L}_k^{UL}(i,t) = \frac{D_k^{UL}(i)}{b \log_2(1 + \frac{p_k^{UL} |(\mathbf{h}_{k,a}^{UL}(i,t))^H \mathbf{v}_k(i,t)|^2}{(\sigma_k^{UL})^2 [((\mathcal{H}_{UL}(i,t))^H \mathcal{H}_{UL}(i,t))^{-1}]_{kk}})}. \tag{4.34}$$

Further substituting (4.34) into constraint (4.30a), yields that downlink latency $\mathcal{L}_k^{DL}(i,t)$ of each mobile user $k \in \mathcal{K}$ within time slot $t \in \mathcal{T}_i$ under DT scene $i \in \mathcal{I}$ cannot exceed $\mathcal{L}_{max} - \mathcal{L}_k^{UL}(i,t) - \mathcal{L}_k^{PRO}(i,t)$. After this, we can decompose problem (4.32) into optimizing $\mathbf{W}(i,t)$ as

$$\min_{\mathbf{W}(i,t)} \sum_{k \in \mathcal{K}} \varpi_k^{\iota}(i) \mathcal{L}_k^{DL}(i,t), \tag{4.35}$$

$$s.t., \text{ (4.23d), (4.31b),}$$

$$\mathcal{L}_k^{DL}(i,t) \le \mathcal{L}_{max} - \mathcal{L}_k^{UL}(i,t) - \mathcal{L}_k^{PRO}(i,t), \forall k \in \mathcal{K}. \tag{4.35b}$$

Similar to problem (4.32), we can derive $\mathbf{W}(i,t)$ following the classic ZF algorithm as

$$\begin{aligned} \mathbf{W}(i,t) &= \mathcal{H}_{DL}(i,t)((\mathcal{H}_{DL}(i,t))^H \mathcal{H}_{DL}(i,t))^{-1} \mathbf{P}^{\frac{1}{2}} \\ &= \tilde{\mathbf{W}}(i,t)\mathbf{P}^{\frac{1}{2}}, \end{aligned} \tag{4.36}$$

where $\mathcal{H}_{DL}(i,t) = \{\mathbf{h}_{1,a}^{DL}(i,t), \ldots, \mathbf{h}_{K,a}^{DL}(i,t)\} \in \mathbb{C}^{M \times K}$ is the matrix of effective downlink channels for all users within time slot $t \in \mathcal{T}_i$ under DT scene $i \in \mathcal{I}$, $\tilde{\mathbf{W}}(i,t) = \mathcal{H}_{DL}(i,t)((\mathcal{H}_{DL}(i,t))^H \mathcal{H}_{DL}(i,t))^{-1}$, and $\mathbf{P}$ is a diagonal matrix whose kth diagonal element is the receive power at mobile user $k \in \mathcal{K}$ within time slot $t \in \mathcal{T}_i$ under DT scene $i \in \mathcal{I}$, i.e., $p_k^{DL}(i,t)$. However, it is worth noting that, $\mathbf{P}$ remains unknown in (4.36), preventing us from determining $\mathbf{W}(i,t)$, and thus we should proceed to solve $\mathbf{P}$.

Again, referring to the classic ZF algorithm, the following constraint should be satisfied:

$$|(\mathbf{h}_{a,k}^{DL}(i,t))^H \mathbf{w}_k(i,t)| = \sqrt{p_k^{DL}(i,t)}, \forall k \in \mathcal{K}. \tag{4.37}$$

Then, by applying constraint (4.37), and substituting (4.13) into problem (4.35), we can convert the minimization problem of (4.35) into a maximization problem, i.e.,

$$\max_{\{p_k^{DL}(i,t)\}} \sum_{k \in \mathcal{K}} \frac{b \log_2(1 + \frac{p_k^{DL}(i,t)}{(\sigma_k^{DL})^2})}{\varsigma E_k(i,t) \varpi_k^{\iota}(i)}, \tag{4.38}$$

$$s.t., \text{ (4.23d),}$$

$$\log_2\left(1 + \frac{p_k^{DL}(i,t)}{(\sigma_k^{DL})^2}\right) \ge \frac{\varsigma E_k(i,t)}{b(\mathcal{L}_{max} - \mathcal{L}_k^{UL}(i,t) - \mathcal{L}_k^{PRO}(i,t))}. \tag{4.38b}$$

The eliminated inter-user interference results in separate subchannels, necessitating the subchannel-by-subchannel power allocation for problem (4.38). To this end, we employ the water-filling algorithm, which is well suited for independent subchannel optimization and can provide a closed-form solution [19, 24]. Accordingly, the closed-form solution of $p_k^{DL}(i,t)$ for each mobile user $k \in \mathcal{K}$ within time slot $t \in \mathcal{T}_i$ under DT scene $i \in \mathcal{I}$ can be obtained as

$$p_k^{DL}(i,t) = \frac{1}{\upsilon_k(i,t)} \max\left\{\frac{1}{\varrho} - \upsilon_k(i,t)(\sigma_k^{DL})^2, \upsilon_k(i,t)p_k^{DL,min}(i,t)\right\}, \tag{4.39}$$

where $\upsilon_k(i,t)$ is the kth diagonal element of $(\tilde{\mathbf{W}}(i,t))^H\tilde{\mathbf{W}}(i,t)$, ϱ is a normalization factor selected for ensuring $\sum_{k\in\mathcal{K}} \max\{\frac{1}{\varrho} - \upsilon_k(\sigma_k^{DL})^2, \upsilon_k p_k^{DL,min}\} = P^{max}$, which can be derived using classic binary search algorithms [30, 31]. Besides, $p_k^{DL,min}(i,t) = (\sigma_k^{DL})^2(2^{\frac{\varsigma E_k(i,t)}{b(\mathcal{L}_{max}-\mathcal{L}_k^{UL}(i,t)-\mathcal{L}_k^{PRO}(i,t))}} - 1)$ is the minimum received power constraint of mobile user $k \in \mathcal{K}$ within time slot $t \in \mathcal{T}_i$ under DT scene $i \in \mathcal{I}$. Finally, by substituting the obtained $p_k^{DL}(i,t), \forall k \in \mathcal{K}$ into (4.36), the transmit beamforming matrix $\mathbf{W}(i,t)$ can be eventually derived. Overall, the ZF-based optimization algorithm for deriving $\kappa(i,t) = \{\mathbf{V}(i,t), \mathbf{W}(i,t)\}$ is summarized in Algorithm 1.

Algorithm 1: ZF-Based Optimization Algorithm

Input: $\vartheta^{gen}(i,t) = \{\mathbf{\Theta}(i,t), \{E_k(i,t)\}_{\forall k\in\mathcal{K}}, \{f_k(i,t)\}_{\forall k\in\mathcal{K}}\}$,
$\mathcal{H}_{UL}(i,t) = \{\mathbf{h}_{1,a}^{UL}(i,t), ..., \mathbf{h}_{K,a}^{UL}(i,t)\}, \mathcal{H}_{DL}(i,t) = \{\mathbf{h}_{1,a}^{DL}(i,t), ..., \mathbf{h}_{K,a}^{DL}(i,t)\}$.

Derive receive beamforming matrix $\mathbf{V}(i,t) = \mathcal{H}_{UL}(i,t)((\mathcal{H}_{UL}(i,t))^H\mathcal{H}_{UL}(i,t))^{-1}$;

Derive transmit beamforming matrix
$\mathbf{W}(i,t) = \mathcal{H}_{DL}(i,t)((\mathcal{H}_{DL}(i,t))^H\mathcal{H}_{DL}(i,t))^{-1}\mathbf{P}^{\frac{1}{2}} = \tilde{\mathbf{W}}(i,t)\mathbf{P}^{\frac{1}{2}}$, $\mathbf{P}$ is a diagonal matrix whose kth diagonal element is the receive power at the mobile user $k \in \mathcal{K}$ within time slot $t \in \mathcal{T}_i$ under DT scene $i \in \mathcal{I}$, i.e., $p_k^{DL}(i,t)$;

For each $k \in \mathcal{K}$, derive $p_k^{DL}(i,t) = \frac{1}{\upsilon_k(i,t)} \max\{\frac{1}{\varrho} - \upsilon_k(i,t)(\sigma_k^{DL})^2, \upsilon_k(i,t)p_k^{DL,min}(i,t)\}$, where $\upsilon_k(i,t)$ is the kth diagonal element of $(\tilde{\mathbf{W}}(i,t))^H\tilde{\mathbf{W}}(i,t)$, ϱ is the solution of $\sum_{k\in\mathcal{K}} \max\{\frac{1}{\varrho} - \upsilon_k(\sigma_k^{DL})^2, \upsilon_k p_k^{DL,min}\} = P^{max}$, derived by classic binary search algorithms [30, 31], $p_k^{DL,min}(i,t) = (\sigma_k^{DL})^2(2^{\frac{\varsigma E_k(i,t)}{b(\mathcal{L}_{max}-\mathcal{L}_k^{UL}(i,t)-\mathcal{L}_k^{PRO}(i,t))}} - 1)$;

Substituting $\{p_k^{DL}(i,t)\}_{\forall k\in\mathcal{K}}$ into $\mathbf{P}$, $\mathbf{W}(i,t)$ can be eventually derived;

Output: $\mathbf{V}(i,t)$ and $\mathbf{W}(i,t)$.

4.4.5 Implementation of PG-ZFO

Here, we introduce the implementation of the proposed PG-ZFO, including its offline training and online execution, as shown in Fig. 4.4.

4.4.5.1 Offline Training

The offline training of PG-ZFO is mainly to train the prompt-guided decision transformer. This training process can be divided into batching and model training. Specifically, we first provide a dataset $\mathcal{D}^{tra} = \{\mathcal{D}(1^{tra}), \mathcal{D}(2^{tra}), ..., \mathcal{D}(|\mathcal{I}^{tra}|)\}$ for offline training the prompt-guided decision transformer, where each $\mathcal{D}(i^{tra}) \in \mathcal{D}^{tra}$ contains E complete decision-making trajectories, also referred to as episodes, and $\mathcal{I}^{tra}$ is the historical DT scene set arisen from the DT model. Each episode $e \in \{1, ..., E\}$ represents the decision-making process for MDP $\mathcal{M}(i^{tra})$ reformulated

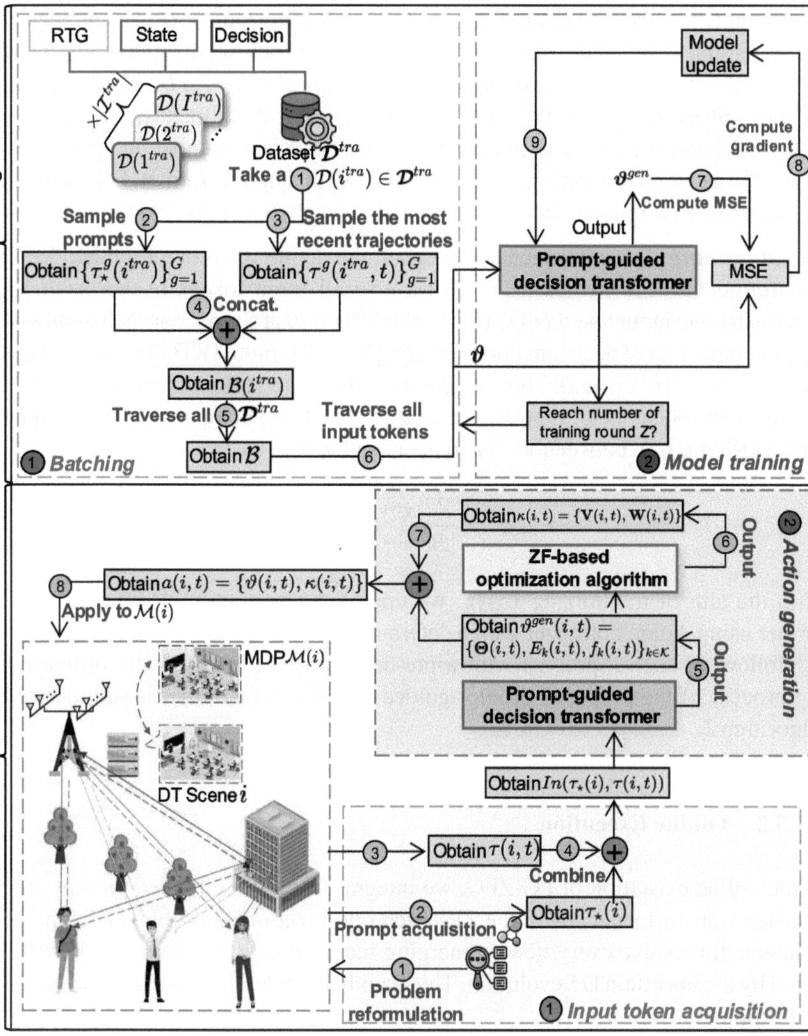

Fig. 4.4 Implementation of the proposed PG-ZFO

from the scene-specific problem $\mathcal{P}(i^{tra})$ under historical DT scene $i^{tra} \in \mathcal{I}^{tra}$ over $\mathcal{T}_{i^{tra}} = \{1, 2, ..., T_{i^{tra}}\}$. Additionally, we further divide each $\mathcal{D}(i^{tra})$ into $\mathcal{D}_\star(i^{tra})$ for sampling the prompt, and $\mathcal{D}_\diamond(i^{tra})$ for sampling the most recent decision-making trajectory.

Batching: For each $\mathcal{M}(i^{tra}), i^{tra} \in \mathcal{I}^{tra}$, we build a minibatch $\mathcal{B}(i^{tra}) = \{In(\tau^g_\star(i^{tra}), \tau^g(i^{tra}, t))\}_{g=1}^G$ includes G pieces input tokens, where every input token $In(\tau^g_\star(i^{tra}), \tau^g(i^{tra}, t)), g \in \{1, ..., G\}$ consists of a prompt $\tau^g_\star(i^{tra})$ sampled from $\mathcal{D}_\star(i^{tra})$, and a most recent decision-making trajectory $\tau^g(i^{tra}, t)$ sampled from $\mathcal{D}_\diamond(i^{tra})$. To increase the training stability and avoid overfitting, for each prompt $\tau_\star^g(i^{tra})$, we first randomly sample U episodes from $\mathcal{D}_\star(i^{tra})$, and then randomly sample a segment from each of them, where each segment is represented as $\tau_\star^{u,g}(i^{tra})$, $u \in \{1, ..., U\}$, with the sequence length of $O = T_\star/U$. Then, we obtain the prompt as $\tau_\star^g(i^{tra}) = (\tau_\star^{1,g}(i^{tra}), ..., \tau_\star^{U,g}(i^{tra}))$. Additionally, for each most recent decision-making trajectory $\tau^g(i^{tra})$, we randomly sample an episode from $\mathcal{D}\diamond(i^{tra})$, and then randomly extract a segment with length L from it. Following this, by combining all minibatches, we can obtain a batch $\mathcal{B} = \{\mathcal{B}(i^{tra}), \forall i^{tra} \in \mathcal{I}^{tra}\}$.

Model training: For convenience, hereafter, we define the prompt-guided decision transformer as $\mathcal{X}_\omega$. We allow $\mathcal{X}_\omega$ to traverse all input tokens in the batch $\mathcal{B}$ by processing one input token $In(\tau_\star^g(i^{tra}), \tau^g(i^{tra})) \in \mathcal{B}$ at a time. For each input token, $\mathcal{X}_\omega$ generates a set of decisions correspondingly to all inputted RTG and state tokens in $In(\tau_\star^g(i^{tra}), \tau^g(i^{tra})) \in \mathcal{B}$. Then, we combine them as a vector of generated decisions for all input tokens in $\mathcal{B}$, represented as $\boldsymbol{\vartheta}^{gen}$. Next, we compute the mean-squared error (MSE) L_{MSE} between $\boldsymbol{\vartheta}^{gen}$ and decisions $\boldsymbol{\vartheta}$ in $\mathcal{B}$ as

$$L_{MSE} = \frac{1}{|\mathcal{B}|} \sum (\boldsymbol{\vartheta}^{gen} - \boldsymbol{\vartheta})^2. \tag{4.40}$$

With the aim of minimizing L_{MSE}, we update the prompt-guided decision transformer using Adam optimizer with a determined learning rate β. We iteratively train $\mathcal{X}_\omega$ following such a process with a pre-determined Z rounds. The offline training process of the proposed prompt-guided decision transformer is summarized in Algorithm 2.

4.4.5.2 Online Execution

In the online execution of PG-ZFO, we integrate the offline-trained prompt-guided decision transformer $\mathcal{X}_\omega$ with the ZF-based optimization algorithm, and apply it to dynamically resolve every newly emerging scene-specific problems $\mathcal{P}(i) \in \boldsymbol{\mathcal{P}}$ triggered by the uncertain DT evolution. This mainly includes the input token acquisition and action generation.

Input token acquisition: For any $\mathcal{P}(i) \in \boldsymbol{\mathcal{P}}$, we first reformulate it as MDP $\mathcal{M}(i) \in \boldsymbol{\mathcal{M}}$. Then, we obtain prompt $\tau_\star(i)$ through a DRL integrated with ZF optimization algorithm based prompt acquisition approach [16, 46, 69], which takes the initial state of the reformulated MDP $\mathcal{M}(i) \in \boldsymbol{\mathcal{M}}$ as the input. We allow it to do decision-making over $T_\star$ time slots under DT scene $i \in \mathcal{I}$. Afterwards, we take current state $s(i, t)$ and the RTG $\hat{r}(i, t) = \hat{R}(i) - \sum_{j=1_\star}^{j=T_\star} r_\star(i, j)$, where current time slot $t = T_\star + 1$, forming the most recent decision-making trajectory $\tau(i, t)$ with sequence length of L. By combing $\tau_\star(i)$ and $\tau(i, t)$, the input token $In(\tau_\star(i), \tau(i, t))$ is obtained.

Action generation: We input $In(\tau_\star(i), \tau(i))$ to PG-ZFO, which use $\mathcal{X}_\omega$ to generate decision $\vartheta^{gen}(i, t) = \{\boldsymbol{\Theta}(i, t), E_k(i, t), f_k(i, t)\}_{k\in\mathcal{K}}$ for $\mathcal{M}(i) \in \boldsymbol{\mathcal{M}}$ within time slot $t \in \mathcal{T}_i$ under DT scene $i \in \mathcal{I}$. Furthermore, we call the designed ZF-based optimization algorithm to derive $\kappa(i, t) = \{\mathbf{V}(i, t), \mathbf{W}(i, t)\}$ by taking $\vartheta^{gen}(i, t)$ as the input. Subsequently, PG-ZFO outputs action $a(i,t) = \{\vartheta(i,t), \kappa(i,t)\}$ for MDP $\mathcal{M}(i) \in \boldsymbol{\mathcal{M}}$, obtaining the new RTG $\hat{r}(i,t+1) = \hat{r}(i,t) - r(i,t)$ and next state $s(i,t+1)$, and update the input by popping the first decision-making trajectory tuples in $\tau(i)$ and adding $(\vartheta^{gen}(i,t), \hat{r}(i,t+1), s(i,t+1))$ at the end. We iteratively conduct such a decision-making process until reaching T_i. The online execution of PG-ZFO is summarized in Algorithm 3.

Note that, the action $a(i,t) = \{\vartheta^{gen}(i,t), \kappa(i,t)\}$ outputted by PG-ZFO corresponds to the decisions for $\mathcal{P}(i) \in \boldsymbol{\mathcal{P}}$ within each time slot $t \in \mathcal{T}_i$ under DT scene $i \in \mathcal{I}$, and thus PG-ZFO indeed solves $\mathcal{P}(i) \in \boldsymbol{\mathcal{P}}$.

Algorithm 2: Offline Training

Input: Dataset $\mathcal{D}^{tra}$, total training steps Z, size of minibatch G, lenaring rate β, initial prompt-guided decision transfomer $\mathcal{X}_\omega$.

for *Each training round* $z = 1$ *to* Z **do**
- **for** *Each MDP* $\mathcal{M}(i^{tra})$, $i^{tra} \in \mathcal{I}^{tra}$ **do**
  - **for** $g = 1$ *to* G **do**
    - Randomly sample U episodes from $\mathcal{D}_\star(i^{tra})$;
    - Randomly sample segments $\tau_\star(i^{tra}, u, g)$ from each sampled episode with length O;
    - Obtain the prompt $\tau_\star(i^{tra}, g)$;
    - Randomly sample an episode from $\mathcal{D}_\diamond(i^{tra})$;
    - Randomly sample a segment with length L from the episode, obtaining the most recent decision-making trajectory $\tau(i^{tra}, t_g, g)$;
    - Obtain the input token $In(\tau_\star(i^{tra}, g), \tau(i^{tra}, t_g, g))$;
  - Get a minibatch $\mathcal{B}(i^{tra}) = \{In(\tau_\star(i^{tra}, g), \tau(i^{tra}, t_g, g))\}_{g=1}^{G}$;
- Get a batch $\mathcal{B} = \{\mathcal{B}(i^{tra}), \forall i^{tra} \in \mathcal{I}^{tra}\}$;
- $\mathbf{a}^{gen} = \mathcal{X}_\omega^z(In(\tau_\star(i^{tra}, g), \tau(i^{tra}, t_g, g))), \forall In(\tau_\star(i^{tra}, g), \tau(i^{tra}, t_g, g)) \in \mathcal{B}$;
- $L_{MSE} = \frac{1}{|\mathcal{B}|}\sum(\boldsymbol{\vartheta}^{gen} - \boldsymbol{\vartheta})^2$;
- $\omega \leftarrow \omega - \beta \nabla_\omega L_{MSE}$;

Output: Offline-trained promot-guided decision transformer $\mathcal{X}_\omega$.

Algorithm 3: Online Execution

Input: Initial state $s(i, 1)$, sequence length of prompt $T_\star$, offline-trained prompt-guided decision transformer $\mathcal{X}_\omega$, frequnecy $\mathcal{F}_i$.

Obtain prompt $\tau_\star(i)$ with sequence length $T_\star$ based on a DRL integrated with ZF optimization algorithm;

Obtain the most recent decision-making trajectory $\tau(i,t)$;

for $t = T_\star + 1$ *to* T_i **do**
- Concatenate $\tau_\star(i)$ with $\tau(i,t)$ as the input token $In(\tau_\star(i), \tau(i,t))$;
- Obtain generated decision $\vartheta(i,t) = \mathcal{X}_\omega^*(In(\tau_\star(i), \tau(i,t)))[-1]$;
- Obtain $\kappa(i,t) = \{\mathbf{V}(i,t), \mathbf{W}(i,t)\}$ by calling the designed ZF-based optimization alogrithm by taking $\vartheta(i,t)$ as the input ;
- Obtain the action $a(i,t) = \{\vartheta(i,t), \kappa(i,t)\}$;
- Apply the action $a(i,t)$ back to $\mathcal{M}(i) \in \boldsymbol{\mathcal{M}}$, attaining the new RTG $\hat{r}(i,t+1)$ and the next state $s(i,t+1)$;
- Update $\tau(i,t) \leftarrow (\tau(i,t), \vartheta(i,t), \hat{r}(i,t+1), s(i,t+1))$;

4.4.6 Computational Complexity Analysis of PG-ZFO

Here, we analyze the computational complexity of PG-ZFO's offline training, followed by the computation complexity of its online execution.

Theorem 4.1 (Complexity of offline training) *The computational complexity of offline training is* $O(Z \cdot |\mathcal{B}| \cdot X \cdot (3(T_\star + L)^2 \cdot d_{embed} + 3(T_\star + L) \cdot d_{embed}^2))$.

Proof The computational complexity of offline training mainly arises from the causal transformer's X stacked identical decoders for processing input tokens. Specifically, for each decoder, the masked multi-head self-attention module processes each input token with computational complexity $O(3(T_\star + L)^2 \cdot d_{embed})$, where $3(T_\star + L)$ signifies the total amount tokens in the input token, and d_{embed} represents the pre-determined embedding dimension of each token after being processed by the corresponding modality-specific trainable linear layer. Subsequently, the feed forward layer processes the input token with computational complexity $O(3(T_\star + L) \cdot d_{embed}^2)$. Since there are X stacked identical decoders, and size of batch $\mathcal{B}$ is $|\mathcal{B}| = G \cdot |\mathcal{I}^{tra}|$, the computational complexity of forward propagation in each training step $z \in \{1, ..., Z\}$ becomes $O(|\mathcal{B}| \cdot X \cdot (3(T_\star + L)^2 \cdot d_{embed} + 3(T_\star + L) \cdot d_{embed}^2))$. Additionally, the computational complexity of backward propagation typically requires approximately twice the computation of the forward propagation, which is $O(2 \cdot |\mathcal{B}| \cdot X \cdot (3(T_\star + L)^2 \cdot d_{embed} + 3(T_\star + L) \cdot d_{embed}^2)) = O(|\mathcal{B}| \cdot X \cdot (3(T_\star + L)^2 \cdot d_{embed} + 3(T_\star + L) \cdot d_{embed}^2))$. Besides, the computational complexity of model update using Adam optimizer is $O(G \cdot d_{embed}^2)$. Therefore, for per training round $z \in \{1, ..., Z\}$, the total computational complexity is $O(|\mathcal{B}| \cdot X \cdot (3(T_\star + L)^2 \cdot d_{embed} + 3(T_\star + L) \cdot d_{embed}^2)) + O(G \cdot d_{embed}^2) = O(|\mathcal{B}| \cdot X \cdot (3(T_\star + L)^2 \cdot d_{embed} + 3(T_\star + L) \cdot d_{embed}^2))$.

Furthermore, since the pre-determined training round is set as Z, the computational complexity of offline training becomes $O(Z \cdot |\mathcal{B}| \cdot X \cdot (3(T_\star + L)^2 \cdot d_{embed} + 3(T_\star + L) \cdot d_{embed}^2))$. □

Theorem 4.2 (Complexity of online execution) *The computational complexity of PG-ZFO in deriving an action* $a(i,t) = \{\boldsymbol{\Theta}(i,t), \{E_k(i,t)\}_{\forall k \in \mathcal{K}}, \{f_k(i,t)\}_{\forall k \in \mathcal{K}}, \mathbf{V}(i,t), \mathbf{W}(i,t)\}$ *during online execution for MDP* $\mathcal{M}(i) \in \boldsymbol{\mathcal{M}}$ *within each time slot* $t \in \mathcal{T}_i$ *is* $O(X \cdot (3(T_\star + L)^2 \cdot d_{embed} + 3(T_\star + L) \cdot d_{embed}^2)) + O((T_\star + L) \cdot (M \cdot K^2 + K^3 + K \cdot \log P^{max}))$.

Proof In the online execution, the computational complexity primarily arises from the forward propagation of the prompt-guided decision transformer and ZF-based optimization algorithm. Specifically, the computational complexity of the forward propagation of the prompt-guided decision transformer is known as $O(X \cdot (3(T_\star + L)^2 \cdot d_{embed} + 3(T_\star + L) \cdot d_{embed}^2))$. Additionally, the computational complexity of ZF-based optimization algorithm includes the complexity of computing $\mathbf{V}(i,t)$, $p_k^{DL}(i,t)$ and $\mathbf{W}(i,t)$ within each time slot $t \in \mathcal{T}_i$ under scene $i \in \mathcal{I}$. Computing $\mathbf{V}(i,t)$ according to (4.33) involves matrix multiplications and a matrix inversion, and thus the computational complexity is $O(M \cdot K^2 + K^3)$. Computing $p_k^{DL}(i,t)$ using watering-filling algorithm leads to the computational complexity $O(M \cdot K^2 + K \cdot \log P^{max})$ [19, 24]. Computing $\mathbf{W}(i, t)$ according to (4.36) involves a matrix multiplication, resulting in the computational complexity is $O(M \cdot K)$. Therefore, the computational complexity of ZF-based optimization algorithm within each time slot $t \in \mathcal{T}_i$ under any scene $i \in \mathcal{I}$ becomes $O(M \cdot K^2 + K^3 + K \cdot \log P^{max})$.

To sum up, the computational complexity of PG-ZFO in deriving an action $a(i, t) = \{\boldsymbol{\Theta}(i, t), \{E_k(i, t)\}_{\forall k \in \mathcal{K}}, \{f_k(i, t)\}_{\forall k \in \mathcal{K}}, \mathbf{V}(i, t), \mathbf{W}(i, t)\}$ for MDP $\mathcal{M}(i) \in \mathcal{M}$ within each time slot $t \in \mathcal{T}_i$ is $O(X \cdot (3(T_\star + L)^2 \cdot d_{embed} + 3(T_\star + L) \cdot d_{embed}^2)) + O((T_\star + L) \cdot (M \cdot K^2 + K^3 + K \cdot \log P^{max}))$. □

4.5 Simulation Results

In this section, we conduct extensive simulations to evaluate the performance of the proposed PG-ZFO in maximizing the sum of all mobile users' QoE in DT interactions under any DT scene, given that the DT model is uncertainly evolved.

4.5.1 Simulation Settings

Consider an RIS-assisted DT interaction system in a 100 m × 100m square area, consisting of a DT server (i.e., base station) with $M = 64$ receive/transmit antennas located at $\mathbf{q_a} = [0\,\text{m}, 0\,\text{m}, 40\,\text{m}]$, an RIS with $N = 16$ passive reflecting elements located at $\mathbf{q_r} = [75\,\text{m}, 100\,\text{m}, 20\,\text{m}]$, and $K = 10$ single-antenna mobile users following a Random-Waypoint mobility pattern [23]. The DT server maintains a DT model that uncertainly evolves in response to the changes of the physical world as depicted in [51]. Table 4.2 lists the values of main simulation parameters, while most of them have also been widely employed in the literature [37, 61, 62]. Besides, we take $|\mathcal{I}^{tra}| = 40$ DT scenes emerged in the DT model as the historical DT scene set $\mathcal{I}^{tra}$. Then, to obtain the dataset $\mathcal{D}^{tra}$, for each corresponding historical scene-specific problem $\mathcal{P}(i^{tra})$, $i^{tra} \in \mathcal{I}^{tra}$, we employ an approach that integrates DRL with ZF-based optimization algorithm [16, 46, 69] to collect $E = 100$ episodes from the reformulated MDP $\mathcal{M}(i^{tra})$, $i^{tra} \in \mathcal{I}^{tra}$, forming $\mathcal{D}(i^{tra}) \in \mathcal{D}^{tra}$. Additionally, for illustration, we select another 2 representative DT scenes to form $\mathcal{I}$, and name them as DT scene 1^{new} and 2^{new}. Particularly, for these new DT scenes, we set the degrees of each mobile user k's, $k \in \mathcal{K}$, attentions to the subjective experience, i.e., $\varpi_k^{\epsilon}(i)$, as a random value in the ranges [0, 0.3] and [0.7, 1], respectively. The main hyperparameters of the proposed PG-ZFO is summarized in Table 4.3. We implement our simulation code using Pytorch and Huggingface Transformers library. All simulations are conducted on a computing platform equipped with two NVIDIA RTX 4090 GPUs, each with 24 GB of memory.

For the comparison purpose, besides the proposed PG-ZFO, the following solution schemes are also simulated as benchmarks.

Table 4.2 Main simulation parameters

Parameter	Value	Parameter	Value
K	10	M	64
N	16	B	2 MHz
ξ	8 Mb	p^{max}	43 dBm
p_k^{UL}	500 mW	C	10 GHz
μ	50 cycle/bit	$D_k(i)$	[0.1, 1.0] MB
$E_k(i)$	[1.0, 2.0]	E_{th}	1.0
ς	1 MB	$\mathcal{L}_{max}$	0.5 s
$(\sigma_k^{UL})^2$	−60 dBm	$(\sigma_k^{DL})^2$	−50 dBm
ρ	−20 dB	$\{G_{k,r}, G_{r,k}\}_{k=1}^K$	8 dB
$G_{r,a}$	6 dB	$G_{a,r}$	7 dB
T_i	100	$T_\star$	12
U	3	L	20

Table 4.3 Main hyperparameters of the proposed PG-ZFO

Parameter	Value	Parameter	Value
G	16	β	1e-4
Learning rate decay weight	1e-4	X	12
d_{embed}	128	Dropout rate	0.1
Transformer activation	ReLU	Optimizer	AdamW

- *Rigid optimization method (ROM)* [21]: ROM tailors a solution policy for a random historical scene-specific problem by DRL, and which is then deployed to solve any new scene-specific problem.
- *Decision transformer without prompt (DF-WP)* [9]: DF-WP is offline trained on the same dataset as the one used in PG-ZFO, and then deployed for online solving any new scene-specific problem. However, it does not consider the augmentation of prompts in either offline training and online execution.

4.5.2 Performance Evaluations

Figure 4.5 illustrates the convergence of the proposed PG-ZFO. The results demonstrate that MSE L_{MSE} continuously decreases as training round increases, eventually converging to a considerably small value. This indicates that PG-ZFO can gradually learn a generalized solution policy applicable across all historical scene-specific problems within $\{\mathcal{P}(i^{tra})\}_{\forall i^{tra} \in \mathcal{I}^{tra}}$. Additionally, we evaluate the performance of the learned solution policy in new scene-specific problem $\mathcal{P}(1^{new})$. The results show that the sum of all mobile users' QoE in DT interactions under DT scene 1^{new} increases and stabilizes when the L_{MSE} converges, implying that the solution policy

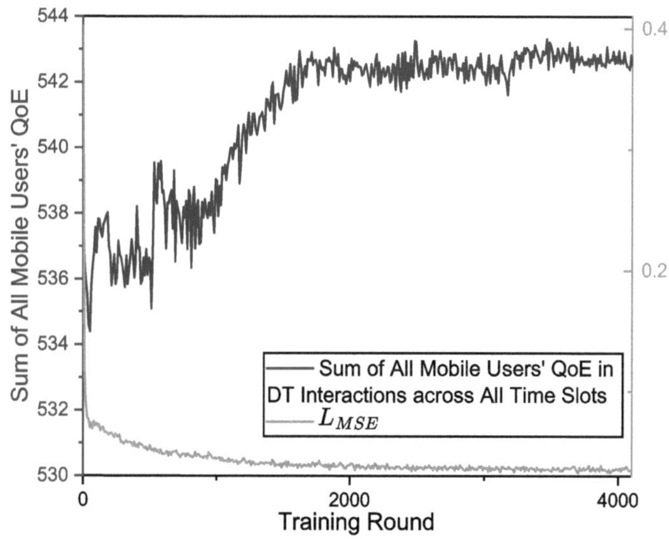

Fig. 4.5 Convergence of the proposed PG-ZFO approach

learned by PG-ZFO can also generalize to unseen scene-specific problems beyond $\{\mathcal{P}(i^{tra})\}_{\forall i^{tra} \in \mathcal{I}^{tra}}$. This can be attributed to the adoption of a well-designed prompt in PG-ZFO, which enriches the representation of scene-specific information, thereby improving its generalization capability.

Figure 4.6 demonstrates the performance of the proposed PG-ZFO in DT scene 1^{new}, under different amounts of decision-making trajectory tuples, i.e., $T_\star$, stored in the prompt. It can be observed that as $T_\star$ increases, the performance of PG-ZFO initially rises, because, with a larger $T_\star$, the prompt can deliver more scene-specific information to effectively guide the proposed PG-ZFO to conduct scene-specific resource allocations. However, its performance becomes relatively stable after $T_\star = 12$, and the reason may be that the processing capacity of PG-ZFO is limited, meaning it can only extract and leverage a limited scene-specific information from the prompt to guide the action generations, regardless of prompt's length. This inspires us to set the appropriate value of $T_\star$ as 12 in PG-ZFO, because a larger $T_\star$ can also result in higher computational complexity in offline training and online execution of PG-ZFO, as analyzed in Sect. 4.4.6.

Figure 4.7 compares the performance of the proposed PG-ZFO with those of ROM and DF-WP in solving new scene-specific problems. It can be observed from Fig. 4.7a that the sum of all mobile users' QoE in DT interactions under new DT scenes increases with the increase of P^{max}. This is because a higher P^{max} allows the DT server to allocate more transmit power to mobile users, thereby accelerating the processing of feedback signals and potentially improving objective performance, as shown in Fig. 4.7b. This, in turn, allows the DT server to configure a higher resolution for mobile users without violating the stringent round-trip latency constraint (4.23e), thereby enhancing the subjective experience, as shown in Fig. 4.7c. Moreover, it is

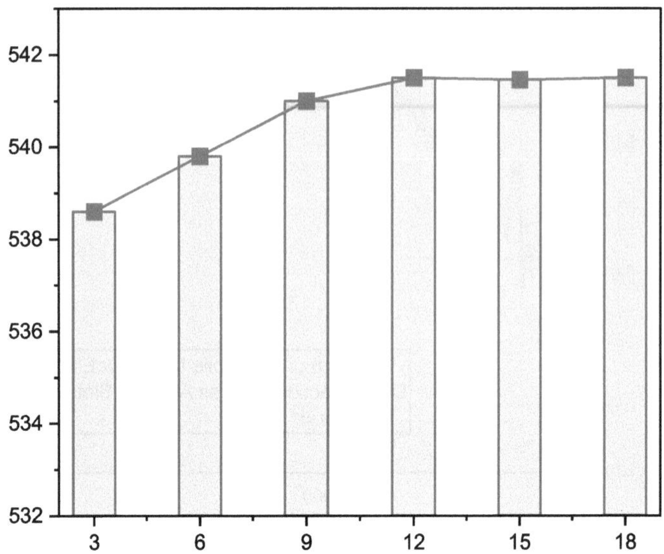

Fig. 4.6 Sum of all mobile users' QoE in DT interactions across all time slots optimized by the proposed PG-ZFO with different amounts of decision-making trajectory tuples $T_\star$ stored in the prompt

shown from Fig. 4.7a that, (i) the proposed PG-ZFO not only exhibits the most significant performance improvement in the sum of all mobile users' QoE as P^{max} increases under both new DT scenes 1^{new} and 2^{new}, but also achieves the best performance compared to ROM and DF-WP; and (ii) DF-WP outperforms ROM in terms of the sum of all mobile users' QoE. This can be explained in twofold. First, in DT scenes 1^{new} and 2^{new}, the prompt can respectively capture the rich scene-specific information of 1^{new} and 2^{new}, including current P^{max} and degrees of each mobile user attentions to the objective and subjective experiences, to guide the PG-ZFO's action generation. For 1^{new}, guided by the prompt, which captures that mobile users engaging in 1^{new} focus more on objective experiences, PG-ZFO allocates more transmit power to mobile users and maintains a relatively constant resolution as P^{max} increases, resulting in a superior performance in terms of objective experience as shown in Fig. 4.7b, while a mediocre one in terms of subjective experiences as shown in Fig. 4.7c. In contrast, in 2^{new}, where users prioritize subjective experiences, guided by the prompt, PG-ZFO not only configures higher resolution but also allocates more transmit power for mobile users, ensuring compliance with the strict round-trip latency constraint (4.23e), resulting in a superior performance in terms of subjective experiences as shown in Fig. 4.7c, while a mediocre one in terms of objective experiences as shown in Fig. 4.7b. Second, despite the forward-looking RTG used in DF-WP, which makes it perform better than ROM, which does not consider generalization, its guidance on action generation is still limited compared to the prompt in PG-ZFO.

Figure 4.8 shows the performance comparison between the proposed PG-ZFO, ROM, and DF-WP in solving new scene-specific problems, under different values

Fig. 4.7 Comparison of the performance of different approaches under various DT scenes with different maximum transmit power P^{max}

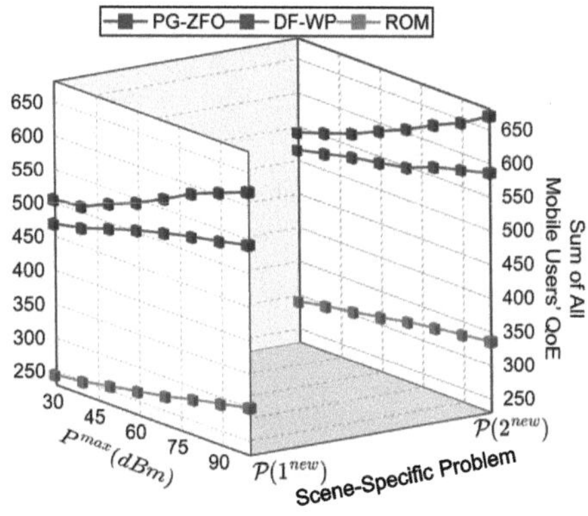

(a) Sum of all mobile users' QoE with respect to P^{max}.

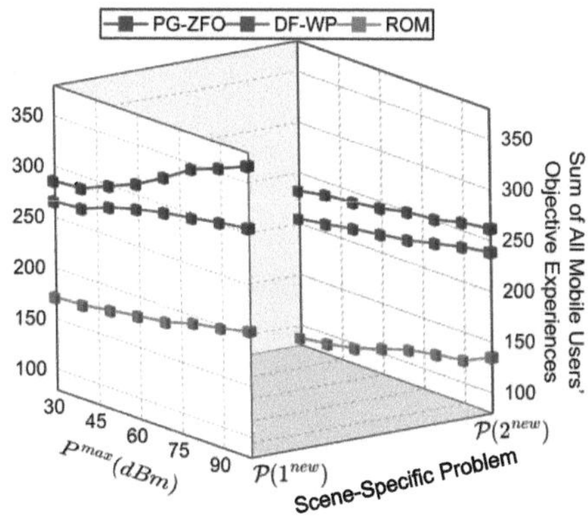

(b) Sum of all mobile users' objective experiences with respect to P^{max}.

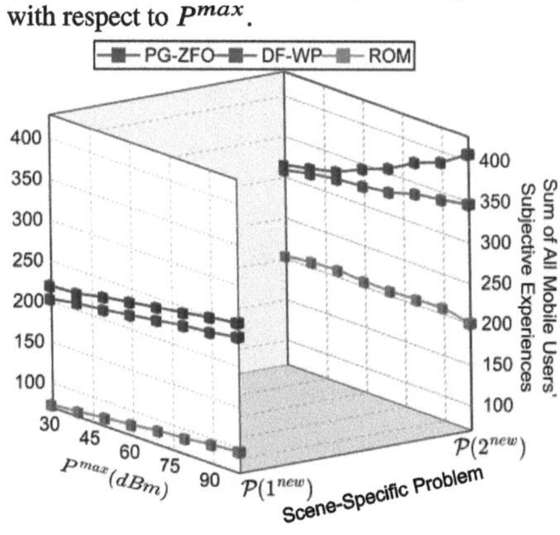

(c) Sum of all mobile users' subjective experiences with respect to P^{max}.

Fig. 4.8 Comparison of the performance of different approaches under various DT scenes with different maximal interaction round-trip latency that each mobile user can tolerate $\mathcal{L}_{max}$

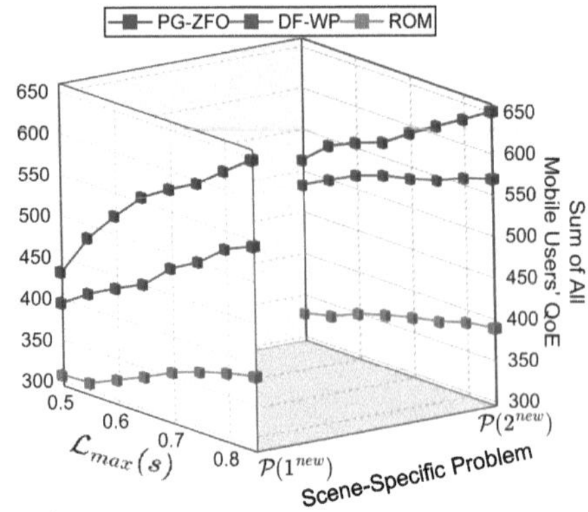

(a) Sum of all mobile users' QoE with respect to $\mathcal{L}_{max}$.

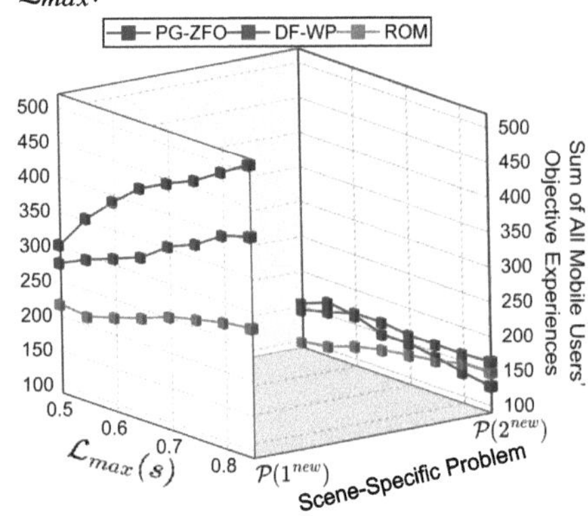

(b) Sum of all mobile users' objective experiences with respect to $\mathcal{L}_{max}$.

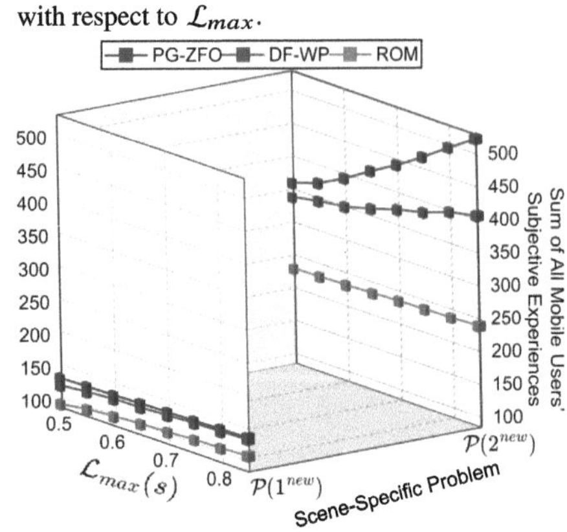

(c) Sum of all mobile users' subjective experiences with respect to $\mathcal{L}_{max}$.

of maximal tolerable round-trip latency $\mathcal{L}_{max}$. It can be observed from Fig. 4.8a that the sum of all mobile users' QoE increases with $\mathcal{L}_{max}$ under both new DT scenes 1^{new} and 2^{new}. This is because that a higher $\mathcal{L}_{max}$, which makes mobile users' to tolerate a larger round-trip latency, allows the DT server to configure a higher feedback signal resolution, and thereby enhancing mobile users' objective and subjective experiences, respectively, which can also be verified in Fig. 4.8b and c. Additionally, it can be observed from Fig. 4.8a that (i) the proposed PG-ZFO demonstrates the most substantial performance improvement under both new DT scenes 1^{new} and 2^{new}, and outperforms benchmark schemes in terms of the sum of all mobile users' QoE; (ii) DF-WP outperforms ROM under both DT scene 1^{new} and 2^{new}. First, on one hand, for DT scene 1^{new}, the prompt, which can capture the information that mobile users focus more on objective experiences and current $\mathcal{L}_{max}$, can guide PG-ZFO to keep the configured resolution to be relatively constant, leading to the significance improvement of each mobile users' objective experiences, as shown in Fig. 4.8b. On the other hand, for DT scene 2^{new}, the prompt captures the information that mobile users focus more on subjective experiences and current $\mathcal{L}_{max}$, which guide PG-ZFO to configure higher resolution, such that benefiting the subjective experiences of mobile users, as shown in Fig. 4.8c. Besides, although increasing $\mathcal{L}_{max}$ may potentially enhance objective performance, the increase in configured resolution can even decrease such performance, as depicted in Fig. 4.8b. Besides, as explained for Fig. 4.7, by capturing limited scene-specific information, RTG enables DF-WP to be generalized better than ROM, but still performs worse than PG-ZFO.

Figure 4.9 illustrates the performance comparison of the proposed PG-ZFO with those of ROM and DF-WP in addressing new scene-specific problems, under various amounts of required computing resource μ for processing 1-bit feedback signals. It can be seen from Fig. 4.9a that the sum of all mobile users' QoE in DT interactions under both DT scenes 1^{new} and 2^{new} decrease with the increase of μ. This is because a larger μ increases the required computing resource to process feedback signals, while the available computing resources of the DT server remain constant, leading to a higher processing latency for each mobile user, and consequently deteriorating their objective experience, as shown in Fig. 4.9b. Furthermore, to avoid the violation of the strict round-trip latency constraint (4.23e), the configured resolution for mobile users may be lowered, in turn leads to the deterioration of their subjective experience, as shown in Fig. 4.9c. Moreover, Fig. 4.9a also shows that the proposed PG-ZFO exhibits the slowest decline in performance, and outperforms all benchmark schemes. The reason is as follows. First, for DT scene 1^{new}, guided by the prompt, which captures the information that mobile users focus more on objective experience and current value of μ, PG-ZFO well balances the objective and subjective experience by sensitively decreasing the configured resolution to mitigate the decline rate of objective experience as μ increases, as shown in Fig. 4.9b. For DT scene 2^{new}, the prompt capturing the information that mobile users focus more on subjective experience, can guide PG-ZFO to slightly decrease the configured resolution while satisfying the round-trip latency constraint (4.23e), such that a relative high subjective experience can be maintained, as shown in Fig. 4.9c.

Fig. 4.9 Comparison of the performance of different approaches under various DT scenes with different amount of required computing resource μ for processing 1-bit feedback signals

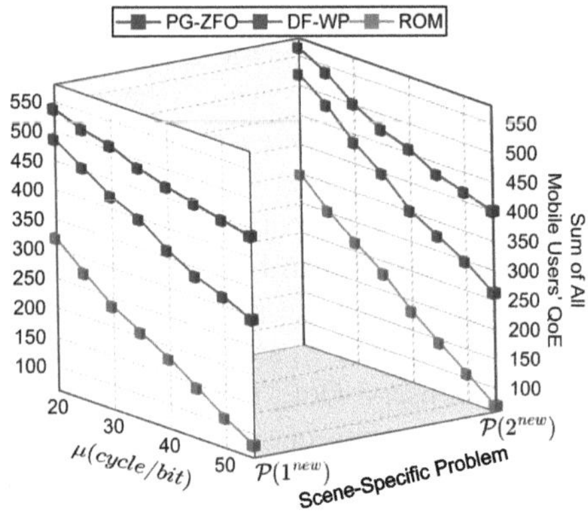

(a) Sum of all mobile users' QoE with respect to μ.

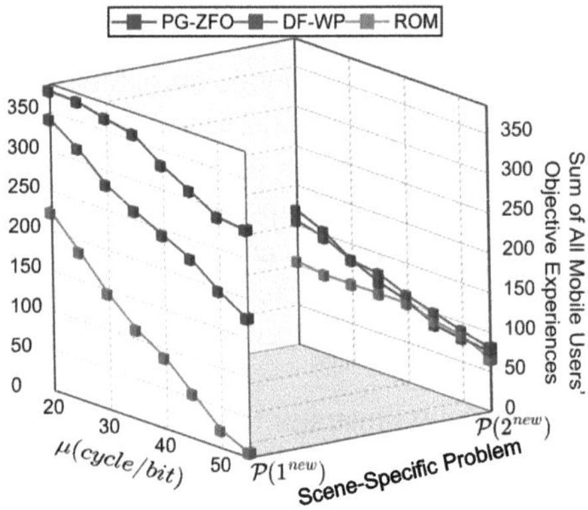

(b) Sum of all mobile users' objective experiences with respect to μ.

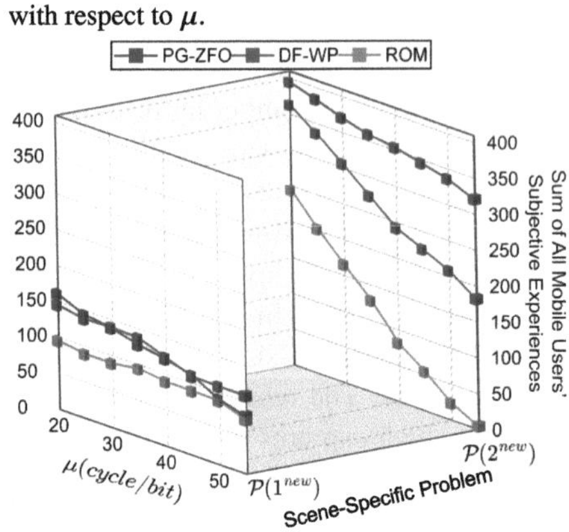

(c) Sum of all mobile users' subjective experiences with respect to μ.

4.6 Summary of the Chapter

In this chapter, we have proposed a novel GAI-aided approach to address the QoE-aware resource allocation for RIS-assisted DT interaction with uncertain evolution. Specifically, considering the distinct features of DT interactions that involve round-trip communications with both uplink and downlink, and with the aim of balancing objective performance and subjective experience, we formulate an optimization problem to maximize the sum of mobile users' QoE by jointly determining phase shift matrix, receive/transmit beamforming matrices, feedback signal rendering resolution and computing resource configuration. Accounting for uncertain evolutions of the DT model and the resulted variations of DT scenes, we extend the problem to a series of scene-specific ones. To obtain a generalized method, we have particularly developed a GAI-aided approach, called PG-ZFO, and explain its implementation in detail. Simulation results have shown the effectiveness and superiority of PG-ZFO over existing approaches.

References

1. Ben-Tal, A., Teboulle, M.: Hidden convexity in some nonconvex quadratically constrained quadratic programming. Math. Program. **72**(1), 51–63 (1996)
2. Brown, T., Mann, B., Ryder, N., Subbiah, M., Kaplan, J.D., Dhariwal, P., Neelakantan, A., Shyam, P., Sastry, G., Askell, A., Agarwal, S., Herbert-Voss, A., Krueger, G., Henighan, T., Child, R., Ramesh, A., Ziegler, D., Wu, J., Winter, C., Hesse, C., Chen, M., Sigler, E., Litwin, M., Gray, S., Chess, B., Clark, J., Berner, C., McCandlish, S., Radford, A., Sutskever, I., Amodei, D.: Language models are few-shot learners. In: Proceedings of NeurIPS, vol. 33, pp. 1877–1901 (2020)
3. Cai, J., Shen, X., Mark, J.W., Alfa, A.S.: Semi-distributed user relaying algorithm for amplify-and-forward wireless relay networks. IEEE Trans. Wirel. Commun. **7**(4), 1348–1357 (2008)
4. Cao, H., Cai, J.: Distributed multiuser computation offloading for cloudlet-based mobile cloud computing: a game-theoretic machine learning approach. IEEE Trans. Veh. Technol. **67**(1), 752–764 (2018)
5. Chen, J., Shi, Y., Yi, C., Du, H., Kang, J., Niyato, D.: Generative AI-driven human digital twin in IoT-healthcare: a comprehensive survey. IEEE Internet Things J. **11**(21), 34749–34773 (2024)
6. Chen, J., Yi, C., Du, H., Niyato, D., Kang, J., Cai, J., Shen, X.: A revolution of personalized healthcare: enabling human digital twin with mobile AIGC. IEEE Netw. **38**(6), 234–242 (2024)
7. Chen, J., Yi, C., Okegbile, S.D., Cai, J., Shen, X.S.: Networking architecture and key supporting technologies for human digital twin in personalized healthcare: a comprehensive survey. IEEE Commun. Surv. Tutor. **26**(1), 706–746 (2024)
8. Chen, J., Yi, C., Wang, R., Zhu, K., Cai, J.: Learning aided joint sensor activation and mobile charging vehicle scheduling for energy-efficient wrsn-based industrial IoT. IEEE Trans. Veh. Technol. **72**(4), 5064–5078 (2023)

9. Chen, L., Lu, K., Rajeswaran, A., Lee, K., Grover, A., Laskin, M., Abbeel, P., Srinivas, A., Mordatch, I.: Decision transformer: reinforcement learning via sequence modeling. In: Proceedings of NeurIPS, vol. 34, pp. 15084–15097 (2021)
10. Chen, L., Zheng, S., Wu, Y., Dai, H.N., Wu, J.: Resource and fairness-aware digital twin service caching and request routing with edge collaboration. IEEE Wirel. Commun. Lett. **12**(11), 1881–1885 (2023)
11. Chen, Q., Li, R., Xu, X., Wu, J., Jiang, H., Qiu, M.: Human-aware dynamic hierarchical network control for distributed Metaverse services. IEEE J. Sel. Areas Commun. **42**(3), 629–642 (2024)
12. Chen, R., Yi, C.: A game-theoretic online optimization for federated digital twin construction via wireless sensing. In: Proceedings of IEEE ICC, Montreal, Canada (2025)
13. Chen, R., Yi, C., Zhou, F., Kang, J., Wu, Y., Niyato, D.: Federated digital twin construction via distributed sensing: a game-theoretic online optimization with overlapping coalitions. IEEE Trans. Mob. Comput., 1–17 (2025)
14. Chen, R., Yi, C., Zhu, K., Cai, J., Chen, B.: A DRL-based hierarchical game for physical layer security with dynamic trilateral coalitions. In: Proceedings of IEEE ICC, pp. 4495–4500 (2023)
15. Chen, R., Yi, C., Zhu, K., Chen, B., Cai, J., Guizani, M.: A three-party hierarchical game for physical layer security aware wireless communications with dynamic trilateral coalitions. IEEE Trans. Wirel. Commun. **23**(5), 4815–4829 (2024)
16. Fujimoto, S., Hoof, H., Meger, D.: Addressing function approximation error in actor-critic methods. In: Proceedings of ICML, pp. 1587–1596. PMLR (2018)
17. Gong, S., Lu, X., Hoang, D.T., Niyato, D., Shu, L., Kim, D.I., Liang, Y.C.: Toward smart wireless communications via intelligent reflecting surfaces: a contemporary survey. IEEE Commun. Surv. Tutor. **22**(4), 2283–2314 (2020)
18. Guc, F., Viola, J., Chen, Y.: Digital twins enabled remote laboratory learning experience for mechatronics education. In: IEEE Proceedings of IEEE DTPI, pp. 242–245 (2021)
19. He, P., Zhao, L., Zhou, S., Niu, Z.: Water-filling: a geometric approach and its application to solve generalized radio resource allocation problems. IEEE Trans. Wirel. Commun. **12**(7), 3637–3647 (2013)
20. Hu, S., Shen, L., Zhang, Y., Chen, Y., Tao, D.: On transforming reinforcement learning with transformers: the development trajectory. IEEE Trans. Pattern Anal. Mach, Intell (2024)
21. Hu, Z., Niu, J., Ren, T., Guizani, M.: Achieving fast environment adaptation of DRL-based computation offloading in mobile edge computing. IEEE Trans. Mob. Comput. **23**(5), 6347–6362 (2024)
22. Jiang, S., Wang, X., Lin, J., Huang, C., Qian, Z., Han, Z.: A delay-oriented joint optimization approach for RIS-assisted MEC-MIMO system. IEEE Trans. Mob. Comput., 1–15 (2024)
23. Johnson, D.B., Maltz, D.A.: Dynamic source routing in Ad hoc wireless networks. Mob. Comput. pp. 153–181 (1996)
24. Kobayashi, M., Caire, G.: An iterative water-filling algorithm for maximum weighted sum-rate of Gaussian MIMO-BC. IEEE J. Sel. Areas Commun. **24**(8), 1640–1646 (2006)
25. Kurma, S., Singh, K., Katwe, M., Mumtaz, S., Li, C.P.: RIS-empowered MEC for URLLC systems with digital-twin-driven architecture. IEEE Trans. Commun. **72**(4), 1983–1997 (2024)
26. Lei Ba, J., Kiros, J.R., Hinton, G.E.: Layer normalization (2016). arXiv e-prints arXiv:1607.06450
27. Li, G., Cai, J.: An online incentive mechanism for collaborative task offloading in mobile edge computing. IEEE Trans. Wirel. Commun. **19**(1), 624–636 (2020)
28. Li, K., Zhu, P., Wang, Y., Zheng, F.C., You, X.: Joint uplink and downlink resource allocation toward energy-efficient transmission for URLLC. IEEE J. Sel. Areas Commun. **41**(7), 2176–2192 (2023)
29. Li, X., Wei, J., Wang, H., Dong, L., Chen, R., Yi, C., Cai, J., Niyato, D., Shen, X.: Towards intelligent transportation with pedestrians and vehicles in-the-loop: a surveillance video-assisted federated digital twin framework. IEEE Netw., 1–1 (2025)

30. Lim, H., Lee, N.: Survey and proposal on binary search algorithms for longest prefix match. IEEE Commun. Surv. Tutor. **14**(3), 681–697 (2011)
31. Lin, A.: Binary search algorithm. WikiJ. Sci. **2**(1), 1–13 (2019)
32. Lin, X., Wu, J., Li, J., Yang, W., Guizani, M.: Stochastic digital-twin service demand with edge response: an incentive-based congestion control approach. IEEE Trans. Mob. Comput. **22**(4), 2402–2416 (2023)
33. Liu, D., Chen, R., Yi, C., Zhang, T., Li, X., Li, J., Wang, R., Zhu, K.: A hierarchical game for physical layer security aware cooperative communications with dynamic interchangeable relaying and jamming. IEEE Trans. Veh. Technol. **74**(1), 968–983 (2025)
34. Liu, D., Chen, R., Zhang, T., Yi, C.: A DRL-based hierarchical game for physical layer security aware cooperative communications. In: Proceedings of IEEE ICCC, pp. 1769–1774 (2024)
35. Liu, X., Deng, Y., Han, C., Renzo, M.D.: Learning-based prediction, rendering and transmission for interactive virtual reality in RIS-assisted terahertz networks. IEEE J. Sel. Areas Commun. **40**(2), 710–724 (2022)
36. Liu, Z., Li, Z., Gong, Y., Wu, Y.C.: RIS-aided cooperative mobile edge computing: computation efficiency maximization via joint uplink and downlink resource allocation. IEEE Trans. Wirel. Commun. pp. 1–1 (2024)
37. Liu, Z., Li, Z., Gong, Y., Wu, Y.C.: RIS-aided cooperative mobile edge computing: computation efficiency maximization via joint uplink and downlink resource allocation. IEEE Trans. Wirel. Commun. **23**(9), 11535–11550 (2024)
38. Luo, C., Jiang, W., Niyato, D., Ding, Z., Li, J., Xiong, Z.: Optimization and DRL-based joint beamforming design for active-RIS enabled cognitive multicast systems. IEEE Trans. Wirel. Commun. **23**(11), 16234–16247 (2024)
39. Ma, Y., Ota, K., Dong, M.: QoE optimization for virtual reality services in multi-RIS-assisted terahertz wireless networks. IEEE J. Sel. Areas Commun. **42**(3), 538–551 (2024)
40. Ma, Y., Ota, K., Dong, M.: QoE optimization for virtual reality services in multi-RIS-assisted terahertz wireless networks. IEEE J. Sel. Areas Commun. **42**(3), 538–551 (2024)
41. Mahdavi Goloujeh, A., Sullivan, A., Magerko, B.: Is it AI or is it me? Understanding users' prompt journey with text-to-image generative AI tools. In: Proceedings of ACM CHI, pp. 1–13 (2024)
42. Mehanna, O., Huang, K., Gopalakrishnan, B., Konar, A., Sidiropoulos, N.D.: Feasible point pursuit and successive approximation of non-convex QCQPs. IEEE Signal Process. Lett. **22**(7), 804–808 (2014)
43. Mihai, S., Yaqoob, M., Hung, D.V., Davis, W., Towakel, P., Raza, M., Karamanoglu, M., Barn, B., Shetve, D., Prasad, R.V., Venkataraman, H., Trestian, R., Nguyen, H.X.: Digital twins: a survey on enabling technologies, challenges, trends and future prospects. IEEE Commun. Surv. Tutor. **24**(4), 2255–2291 (2022)
44. Okegbile, S.D., Cai, J., Yi, C., Niyato, D.: Human digital twin for personalized healthcare: vision, architecture and future directions. IEEE Netw. **37**(2), 262–269 (2023)
45. Park, Y.M., Hassan, S.S., Tun, Y.K., Han, Z., Hong, C.S.: Joint trajectory and resource optimization of MEC-assisted UAVs in sub-THz networks: a resources-based multi-agent proximal policy optimization DRL with attention mechanism. IEEE Trans. Veh. Technol. **73**(2), 2003–2016 (2024)
46. Peel, C.B., Hochwald, B.M., Swindlehurst, A.L.: A vector-perturbation technique for near-capacity multiantenna multiuser communication–Part I: channel inversion and regularization. IEEE Trans. Commun. **53**(1), 195–202 (2005)
47. Qi, K., Wu, Q., Fan, P., Cheng, N., Chen, W., Letaief, K.B.: Reconfigurable-intelligent-surface-aided vehicular edge computing: joint phase-shift optimization and multiuser power allocation. IEEE Internet Things J. **12**(1), 764–777 (2025)
48. Rajcic, N., Llano Rodriguez, M.T., McCormack, J.: Towards a diffractive analysis of prompt-based generative AI. In: Proceedings of ACM CHI, pp. 1–15 (2024)
49. Ramesh, A., Pavlov, M., Goh, G., Gray, S., Voss, C., Radford, A., Chen, M., Sutskever, I.: Zero-shot text-to-image generation. In: Proceedings of ICML, pp. 8821–8831 (2021)

50. Reichl, P., Tuffin, B., Schatz, R.: Logarithmic laws in service quality perception: where microeconomics meets psychophysics and quality of experience. Telecommun. Syst. **52**, 587–600 (2013)
51. Shuguang, L., Lin, B.: Holographic classroom based on digital twin and its application prospect. In: Proceedings of IEEE ICECE, pp. 122–126 (2020)
52. Wang, L., Wang, K., Pan, C., Aslam, N.: Joint trajectory and passive beamforming design for intelligent reflecting surface-aided UAV communications: a deep reinforcement learning approach. IEEE Trans. Mob. Comput. **22**(11), 6543–6553 (2023)
53. Wang, Z., Huang, Y., Song, D., Ma, L., Zhang, T.: Promptcharm: text-to-image generation through multi-modal prompting and refinement. In: Proceedings of ACM CHI, pp. 1–21 (2024)
54. Wu, J., Chen, R., Yi, C.: A DRL-based deviation-aware federated digital twin construction over wireless edge network. In: International Conference on Wireless Artificial Intelligent Computing Systems and Applications (WASA), pp. 196–208. Springer (2025)
55. Xu, J., Ai, B., Wu, L., Zhang, Y., Wang, W., Li, H.: Deep reinforcement learning for computation rate maximization in RIS-enabled mobile edge computing. IEEE Trans. Veh. Technol. **73**(7), 10862–10866 (2024)
56. Yang, Y., Shi, Y., Chen, R., Yi, C., Kang, J.: Online optimization of edge empowered human digital twin deployment and task offloading. In: Proceedings of IEEE ICCC, pp. 856–861 (2024)
57. Yang, Y., Shi, Y., Yi, C., Cai, J., Kang, J., Niyato, D., Shen, X.: Dynamic human digital twin deployment at the edge for task execution: a two-timescale accuracy-aware online optimization. IEEE Trans. Mob. Comput. **23**(12), 12262–12279 (2024)
58. Yi, C., Cai, J., Su, Z.: A multi-user mobile computation offloading and transmission scheduling mechanism for delay-sensitive applications. IEEE Trans. Mob. Comput. **19**(1), 29–43 (2020)
59. Yi, C., Huang, S., Cai, J.: An incentive mechanism integrating joint power, channel and link management for social-aware D2D content sharing and proactive caching. IEEE Trans. Mob. Comput. **17**(4), 789–802 (2018)
60. Yu, X., Yang, Y., Chen, R., Yi, C.: Online optimization of edge vehicle digital twin migration with adaptive mobility prediction. In: Proceedings of IEEE VTC, Chengdu, China (2025)
61. Zhai, L., Zou, Y., Zhu, J., Jiang, Y.: RIS-assisted UAV-enabled wireless powered communications: system modeling and optimization. IEEE Trans. Wirel. Commun. **23**(5), 5094–5108 (2024)
62. Zhai, L., Zou, Y., Zhu, J., Jiang, Y.: RIS-assisted UAV-enabled wireless powered communications: system modeling and optimization. IEEE Trans. Wirel. Commun. **23**(5), 5094–5108 (2024)
63. Zhang, G., Lu, Y., Li, M., Zhong, Z., Quek, T.Q.: Joint uplink and downlink robust transmission for cell-free networks. IEEE Trans. Wirel. Commun. **23**(10), 15308–15321 (2024)
64. Zhang, H., Nie, J., Yu, Y., Xiong, Z., Jiang, W., Niyato, D.: Performance analysis for STAR-RIS assisted SWIPT system over Rayleigh fading channel. In: Proceedings of IEEE ICC, pp. 1468–1473. IEEE (2023)
65. Zhang, J., Li, J., Wang, Z., Shi, L., Jin, S., Chen, W., Poor, H.V.: Decision transformer for wireless communications: a new paradigm of resource management (2024). arXiv preprint arXiv:2404.05199
66. Zhang, W., Wen, Z., Du, C., Jiang, Y., Zhou, B.: RIS-assisted self-interference mitigation for in-band full-duplex transceivers. IEEE Trans. Commun. **71**(9), 5444–5454 (2023)
67. Zhang, Y., Wang, P., Wang, Q., Zhang, H., Xu, L., Sun, W., Wang, B.: Evolutionary game-based adaptive DT association and transfer for wireless computing power networks. IEEE Trans. Green Commun. Netw. pp. 1–1 (2024)
68. Zhao, X., Liu, H., Gong, S., Ju, X., Xing, C., Zhao, N.: Dual-functional MIMO beamforming optimization for RIS-aided integrated sensing and communication. IEEE Trans. Commun. **72**(9), 5411–5427 (2024). https://doi.org/10.1109/TCOMM.2024.3384932

69. Zhi, K., Pan, C., Ren, H., Wang, K.: Ergodic rate analysis of reconfigurable intelligent surface-aided massive MIMO systems with ZF detectors. IEEE Commun. Lett. **26**(2), 264–268 (2021)
70. Zhou, H., Chen, R., Yi, C., Li, J., Cai, J.: A three-party repeated coalition formation game for pls in wireless communications with IRSs. In: Proceedings of IEEE WCNC, pp. 1–6 (2024)
71. Zhou, H., Chen, R., Yi, C., Zhang, J., Kang, J., Cai, J., Guizani, M.: A repeated coalition formation game for physical layer security aware wireless communications with third-party intelligent reflecting surfaces. IEEE Trans. Wirel. Commun. pp. 1–1 (2025)

Chapter 5
Implementation of Digital Twin in Vehicular Network Applications

5.1 Introduction

Intelligent transportation systems (ITSs) are envisioned to utilize information and communication technologies to enhance road safety and traffic management. Since pedestrians and vehicles are the natural and primary participants in transportation systems, incorporating both entities into real-time decision-making and control processes becomes necessary for ITSs, giving rise to the concept of pedestrians and vehicles in-the-loop [23, 43, 60]. Such an "in-the-loop" scheme requires to continuously collect and process data from various sources, dynamically adjusting traffic operations based on real-time pedestrian-vehicle interactions, with both vehicles and pedestrians acting as active participants and decision-makers in the traffic management process. Compared to traditional ITSs, which primarily focus on vehicle modeling and often neglect pedestrians and their impacts [19, 38, 51], ITSs with pedestrians and vehicles in-the-loop can model and analyze the mutual influence between vehicles and pedestrians to provide safer, more efficient, and human-satisfying traffic management. However, enabling ITS applications with pedestrians and vehicles in-the-loop is non-trivial. It requires a cohesive system to characterize real-world traffic environments and conduct behavioral simulations of pedestrians and vehicles.

Recently, digital twin (DT) has been increasingly recognized as a powerful tool for building ITSs, offering real-time virtual representations of traffic operations to support data-driven decision-making and optimize traffic management [15, 27, 31]. The federated digital twin (FDT), which can be seen as an advanced form of DT [10, 19, 51], allows the integration of geographically dispersed DTs into a unified framework. FDT offers significant potential for ITSs to achieve pedestrians and vehicles in-the-loop, even when traffic participants, environments, and infrastructures

C. Yi et al., *Self-Evolving Digital Twin Over Wireless Networks*, Wireless Networks,
https://doi.org/10.1007/978-3-032-08120-9_5

are spread across different regions, each with numerous features and data sources. In addition, FDT allows data processing to be distributed, and thus can improve scalability and reduce latency. Furthermore, FDT enhances ITSs' ability to predict future events using real-time data updates from multiple sources. All these capabilities are vital for proactive decision-making, such as pedestrian crossing safety and traffic signal optimization over a large area with multiple correlated intersections.

4D traffic scene generation creates realistic three-dimensional representations of transportation elements, such as roads, vehicles, pedestrians, and infrastructure, while incorporating the fourth dimension of time to capture ongoing changes and movements. Real-time processing and interactivity with dynamic environments are critical for managing complex transportation systems, wherein numerous traffic participants operate independently under traffic rules and frequently interact. Offline and non-interactive frameworks [14, 54] cannot adapt to real-time changes, provide dynamic feedback, or integrate with ITSs. Besides, traditional AI models are restricted to processing specific data types and performing predefined tasks in specific environments. As a result, they cannot meet the dynamic and adaptive requirements of 4D traffic scene generation, which requires multi-modal data processing and continuous adaptation to dynamic traffic conditions. The development of artificial intelligence-generated content (AIGC) offers a promising solution [7, 17]. AIGC uses advanced generative models to create traffic scenes and enhances interactivity by generating new content in response to user inputs. AIGC is thus essential for realizing the full potential of real-time interactive 4D traffic scene generation.

AIGC-driven real-time interactive 4D traffic scene generation offers vital benefits to ITSs, including greater adaptability to dynamic environments, improved decision-making in complex situations, and increased interactivity with natural user instructions. For example, an AIGC-driven real-time interactive 4D traffic scene generation system in autonomous driving allows vehicles to interact effectively with their surroundings and perform safe driving operations by continuously updating traffic scenes. These advantages make the AIGC-driven system essential for advancing ITSs, leading to safer, more efficient, and more responsive traffic management. To fully harness AIGC's potential, we leverages vehicular networks, which facilitate continuous data exchange between vehicles and infrastructure elements like traffic lights and roadside sensors. However, using AIGC to create real-time interactive 4D traffic scenes over vehicular networks with limited computational resources is very challenging. Although text-to-video models such as Sora [3] can generate high-fidelity, coherent videos from user-provided descriptions, they are not suitable for real-time interactive 4D traffic scene generation because (i) they require significant computational power and time to produce high-quality videos and only generate static outputs based on text inputs, preventing real-time interactivity; and (ii) due to technical limitations, existing text-to-video models cannot maintain the long-term temporal coherence of dynamic traffic elements. To address these severe limitations of existing AIGC models, this chapter proposes a novel framework for AIGC-driven real-time interactive 4D traffic scene generation over vehicular networks, designed to deliver low-latency and interactive traffic services.

5.2 Related Work

In recent years, several DT frameworks have been developed to advance ITS applications [19, 38, 41, 43, 51]. For example, Wang et al. [41] proposed a cloud-edge-device collaborative DT framework for mobility services, but it did not provide implementation details. A DT co-simulation framework for pedestrians and vehicles was introduced in [43], but it lacked twin agent modeling for these traffic participants. Khan et al. [19] and Yu et al. [51] proposed two general FDT architectures for ITSs; however, neither included a technical system implementation. Yang et al. [38] presented an FDT-enabled collision warning framework for autonomous driving, focusing on a semi-asynchronous federated learning scheme to reduce training delays. Despite these contributions, most prior works focus on high-level architectural designs or isolated applications and share several critical limitations: (i) lack of pedestrian-vehicle interaction modeling, existing DT solutions primarily emphasize vehicle modeling while overlooking pedestrian-vehicle interactions and their impacts on traffic dynamics; (ii) high communication costs and privacy risks, centralized ITS architectures require processing large volumes of surveillance video data, leading to excessive bandwidth usage, latency, and privacy concerns; (iii) absence of practical FDT implementations. No existing framework integrates pedestrian-vehicle interaction modeling with real-time FDT representations in complex traffic environments. These gaps highlight the urgent need for a holistic framework that addresses these intertwined challenges.

To address these limitations, in this chapter, we propose a surveillance video-assisted FDT implementation framework (SV-FDT) for ITSs with pedestrians and vehicles in-the-loop. The proposed SV-FDT is built on a cloud-edge-end collaborative architecture and introduces the agent concept to construct pedestrian-vehicle interaction models. The framework integrates semantic segmentation technology [55] for processing surveillance videos, which allows pix-level object recognition and offers granular insights into video data. SV-FDT can achieve timely video acquisition and data integration from multiple sources, precise modeling of pedestrian-vehicle twin agents at different positions, and seamless aggregation of local DT models across different regions. To our best knowledge, SV-FDT is the first DT implementation framework to enable ITSs with pedestrian and vehicle in-the-loop. Our key contributions are summarized as follows:

- We propose SV-FDT, a novel cloud-edge-end collaborative framework for implementing ITSs with pedestrians and vehicles in-the-loop. Its architecture includes (i) an end layer that collects surveillance videos from widespread traffic surveillance cameras; (ii) an edge layer that performs visual understanding, agent-based interaction modeling, and LDTS creation in local regions; and (iii) a cloud layer that integrates LDTSs across geographically dispersed regions to construct a global DT model in real-time. To unlock its full potential, SV-FDT seamlessly integrates algorithms for semantic segmentation, semantic-to-code transformation, and agent-based modeling.

- We analyze key design requirements and challenges for semantic segmentation-based FDT construction in SV-FDT, offering practical guidance for deploying FDTs via traffic video surveillance. We also highlight potential but promising solutions to these challenges, inspiring effective SV-FDT implementation in dynamic transportation.
- We validate SV-FDT through a case study in the CARLA simulation environment to optimize traffic light control in traffic management. Our results show that SV-FDT outperforms traditional terminal-server frameworks in terms of mirroring delay, recognition accuracy, and subjective evaluations.
- We summarize the open challenges in this field and propose future research directions.

Meanwhile, recent advancements in 4D scene generation and AIGC have significantly impacted ITSs [37, 39, 56]. In 4D scene generation, models that combine image, multiview, and video diffusion techniques have been developed to create dynamic scenes [56]. Neural Radiance Fields (NeRF) generate photorealistic scenes as continuous volumetric fields, with Dynamic NeRF incorporating temporal elements to capture object deformations and evolving interactions [39]. Similarly, D-3DGS employs deformable 3D Gaussian Splats to create dynamic scenes with improved photorealism and structural integrity. Savkin et al. [37] introduced an unsupervised neural network architecture for realistic traffic scene generation using synthetic 3D scene graphs. As aforementioned, these text-to-video and text-to-image models take significant computational time and cannot maintain continuous temporal coherence. Thus, they cannot be efficiently applied to generating real-time interactive 4D traffic scenes in ITSs.

In parallel, AIGC-based ITS applications have also emerged [30, 32, 40, 53]. Zhang et al. [53] proposed a cloud-edge-terminal collaborative architecture for autonomous driving, leveraging AIGC for system design and resource management. Masri et al. [32] employed GPT-4o-mini to optimize traffic flow and reduce wait times at urban intersections. Wang et al. [40] integrated large language models (LLMs) into traffic signal control systems, while Manas et al. [30] developed a framework to translate natural language traffic rules into metric temporal logic for autonomous vehicles. However, the potential of AIGC to enable real-time, interactive 4D traffic scene generation over vehicular networks remains unexplored.

To fill this research gap, this chapter further proposes a novel AIGC-driven real-time interactive 4D traffic scene generation framework over vehicular networks. Our key contributions include the following:

- We introduce a novel cloud-edge-terminal collaborative system architecture that supports text-to-code model-driven real-time 4D traffic scenes over vehicular networks. This architecture covers the entire AIGC lifecycle, from data collection and management to model pretraining, fine-tuning, and inference, providing valuable insights for ITSs implementation.
- We conduct an in-depth analysis of the critical design requirements and challenges for deploying real-time, interactive 4D traffic scene generation systems using text-

to-code models. This analysis offers a roadmap for addressing these challenges and ensuring effective implementation in dynamic transportation environments.
- We detail the practical implementation of a real-time, interactive 4D traffic scene generation system based on text-to-code models. This system demonstrates its ability to generate interactive 4D scenes in real time. To our best knowledge, this is the first instance of successfully implementing real-time interactive 4D traffic scene generation. We evaluate the framework's performance by comparing traditional and microservice-based frameworks in terms of inference time and execution accuracy.
- We outline future research directions by highlighting key open questions to inspire further exploration and innovation in this field.

5.3 A Surveillance Video-Assisted Federated DT Framework for ITSs

5.3.1 System Model and Design Goals

5.3.1.1 System Architecture

The proposed SV-FDT system architecture consists of three layers: end, edge, and cloud, as illustrated in Fig. 5.1.

End Layer: This layer consists of terminal devices such as surveillance cameras, radars, and in-vehicle sensors. Its primary function is to collect large-scale, real-time traffic data, particularly surveillance videos. For instance, high-definition (HD) surveillance cameras capture detailed visual data on traffic scenarios, such as vehicle movement, pedestrian crossings, and road conditions. Radars detect and track vehicles and pedestrians, providing information on their speed, location, and movement. In-vehicle sensors, e.g., GPS and ultrasonic sensors, contribute precise real-time localization and movement data. In addition to real-time data collection, these devices with integrated processing modules perform data fusion. Combining information (e.g., the locations and speeds of vehicles and pedestrians) from multiple sources can reduce uncertainties from solely relying on a single data source.

Edge Layer: At the edge layer, edge nodes, equipped with GPUs or NPUs, are responsible for data preprocessing tasks, including (i) redundancy elimination to video data uploaded by terminal devices for abundant storage and bandwidth; (ii) data aggregation to obtain unified data from heterogeneous sources, e.g., traffic cameras, radars, and in-vehicle sensors, by combining radar measurements and in-vehicle sensor data with the visual analysis results; and (iii) synchronization of temporal data to minimize latency discrepancies [7]. More importantly, the edge nodes should conduct semantic segmentation to classify each pixel within each frame of surveillance video, enabling effective identification of pedestrians, vehicles, road markings, traf-

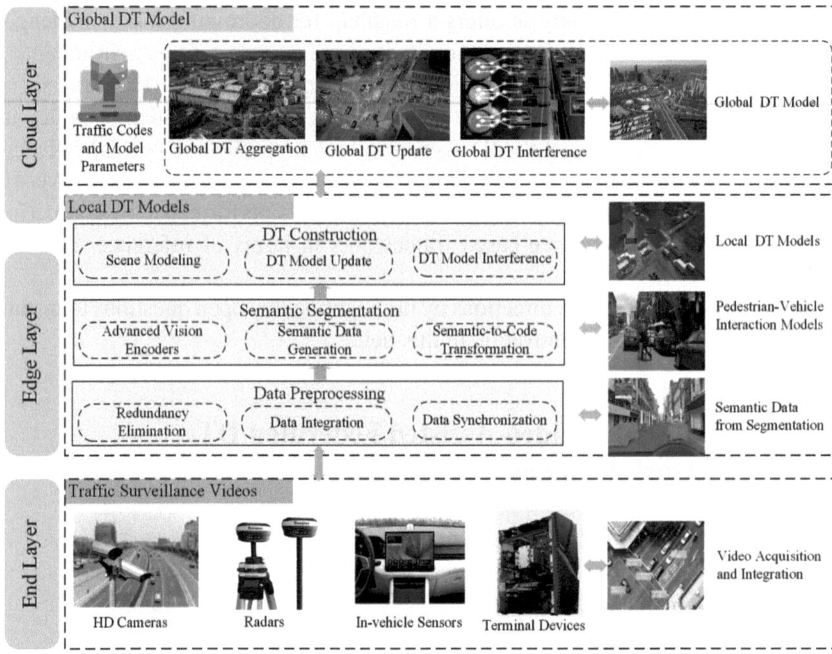

Fig. 5.1 The overall architecture of the proposed SV-FDT system. SV-FDT consists of three layers: the end layer contains multiple terminal devices; the edge layer handles tasks for data preprocessing, semantic segmentation, and local DT construction; and the cloud layer constructs the global model and conducts model inferences for ITS applications

fic signs, barriers, and weather conditions. Powered by advanced vision encoders, it extracts detailed visual features for precise pixel-level classification, continuously generating structured semantic data that accurately represents current traffic scenes. After that, the edge layer adopts a semantic-to-code transformation module that converts semantic data into machine-readable traffic codes, enabling their direct use in traffic simulation systems and facilitating the DT creation and update of physical environments by edge nodes. Meanwhile, the edge nodes transmit traffic codes and model parameters of twin agents to the cloud layer for aggregation to maintain spatiotemporal consistency between local and global DT models. Decision-making and inference results are returned to end-layer terminal devices, enabling adaptive signal control, path optimization, and personalized guidance to improve pedestrian safety and vehicle efficiency.

Cloud Layer: The cloud layer aggregates local DT models from distributed edge devices, creating an up-to-date global DT representation using traffic codes and agent-based model parameters across the transportation network. The global DT model continuously adapts to updates from local models, capturing new patterns in pedestrian-vehicle interactions. Only traffic codes and model parameters are transmitted to protect data privacy, safeguarding sensitive information about pedestrians and vehicles. After updating the global model, the cloud layer distributes global DT model parameters to the edge layer for synchronized updates of local models, enhancing decision-making accuracy.

5.3.1.2 Example Scenarios

Here, we present three typical application scenarios of the proposed SV-FDT framework for ITSs with pedestrians and vehicles in-the-loop.

Extreme Pedestrian-Vehicle Flow Testbed In-the-Loop: SV-FDT can effectively implement ITSs with pedestrians and vehicles in-the-loop by providing a dynamic testbed to manage their interactions. The framework can support various automated driving levels, from Level 0 (fully manual) to Level 5 (fully autonomous) [36]. In each scenario, SV-FDT simulates real-world traffic challenges, such as sudden pattern changes and adverse weather conditions, and adapts by incorporating real-time data from diverse sources. This integration allows the testbed to reflect the varying capabilities of autonomous systems across different levels. By leveraging predictive analytics and collaborative decision-making, FDTs proactively identify and manage potential pedestrian-vehicle conflicts, unlike traditional testbeds that can only react after these situations occur.

Emergency and Disaster Management: Integrating pedestrians and vehicles into SV-FDT can strengthen emergency and disaster management within ITSs. Traditional DT systems without in-the-loop characteristics rely on historical traffic data or static models for delayed emergency responses. Unlike these systems, SV-FDT can continuously monitor and simulate real-world traffic scenarios, and adapt vehicles to unexpected traffic events and situations, including traffic incidents, road blockages, vehicle breakdowns, and sudden pedestrian crowds. Additionally, the system uses predictive models to anticipate emergencies, enabling the accurate assessment of traffic management strategies such as rerouting vehicles and safety notification of pedestrians. The in-the-loop characteristic means that the behaviors of pedestrians and vehicles are integrated into the system's decision-making processes, allowing for proactive rather than reactive management.

Customized Navigation and Travel Guidance: Without in-the-loop FDTs, conventional navigation guidance systems rely on static or historical traffic data. However, they do not account for real-time conditions such as pedestrian activities, sudden traffic disruptions, or evolving road conditions. SV-FDT can be instrumental in enabling personalized navigation systems that cater to the specific needs of both drivers and pedestrians. By integrating data from various sources, such as surveillance cameras and in-vehicle sensors, FDTs create a dynamic, user-centered environment that adapts to individual preferences. For example, commercial drivers can receive guidance to avoid narrow streets or select routes suitable for heavy vehicles. Likewise, family drivers are provided with optimized routes that prioritize safety and efficiency. This adaptability allows FDTs to provide personalized, context-aware travel guidance that enhances the user experience and facilitates seamless interactions between pedestrians and vehicles in ITSs.

5.3.1.3 Key Design Goals

To establish a cloud-edge-end collaborative framework for real-time FDT construction in ITSs [9, 10, 45], we identify the following key design requirements and challenges.

Ultra-Reliable and Agile Global DT Model Aggregation with Pedestrian-Vehicle Interactions: Real-time pedestrian-vehicle interactions require ultra-reliable data transmission to ensure safety and enable quick adjustments to traffic signals and vehicle routing. Therefore, achieving "ultra-reliability" and "agility" in aggregating global DT models from local data is paramount.

- **Timely Video Acquisition and Integration with Precise Co-camera Control**: Real-time video data collection and integration from potentially heterogeneous terminal cameras present challenges in synchronization and coordination [48]. Therefore, a distributed network protocol is essential for timely video acquisition and integration, ensuring that video data is aligned temporally and processed without delays for real-time and accurate DT modeling [8]. Coordination techniques, such as distributed scheduling and synchronization [4], facilitate seamless communication and data sharing among devices, ensuring efficient real-time video data aggregation from multiple cameras. To ensure smooth HD streaming, the end devices, typically the surveillance camera, must achieve a minimum video transmission rate of 10 Mbps and a frame rate of 30 frames per second, supporting H.264 or H.265 video coding [59]. Besides, cameras can be retrofitted with GPU-powered hardware modules to improve efficiency by using lightweight semantic segmentation algorithms to extract granular motion features of traffic elements (e.g., speeds, locations, and directions). Driven by these technologies, co-camera control enables multiple cameras to collaboratively capture videos from various angles and locations, ensuring comprehensive data integrity.
- **Multi-source Data Processing and Distributed Data Synchronization with Privacy Protection**: Leveraging multi-modal data collected from diverse sources in ITSs, e.g., HD cameras and radars, provides valuable insights into ITSs' full feature spaces, including the appearance, locations, directions, and speed of both pedestrians and vehicles. Consolidating such information requires advanced multi-modal data processing and distributed synchronization techniques to create unified feature profiles. This data processing and synchronization process must occur efficiently, ensuring quick updates to the global DT model. However, collecting and transmitting multi-modal data across numerous sources may face significant privacy challenges. From a cross-layer security perspective, privacy-aware access

control and signal protection in both the physical and data link layer, user indistinguishability enhancement, cross-device authentication, and customized packet encryption in upper layers are essential. Advanced technologies, such as physical layer security [11, 12, 25, 26, 57, 58], local differential privacy, zero-trust architecture, and blockchain, can effectively mitigate these risks.

- **Elastic Resource Orchestration for Fine-Grained DT Visualization under Traffic Dynamics**: Edge servers play a critical role in constructing DTs with low-latency requirements yet face significant challenges. High latency can hinder the real-time monitoring of pedestrians and vehicles, leading to uncoordinated interactions and safety risks. Therefore, an integrated approach to task scheduling, communication spectrum management, computing resources, and storage allocation becomes essential [13]. Improving data transmission reliability requires employing ultra-reliable, low-latency communication (URLLC) technologies such as time-sensitive networks [2] and mobile edge computing. These technologies support DT services, including immersive virtual reality and collaborative autonomous driving. For instance, immersive virtual reality demands high-rate video delivery from end to edge, requiring 99.999999% reliability and a latency of no more than 1 ms [20].

Customized and Sustainable Multi-agent Maintenance: To enable customized pedestrian-vehicle interactions in ITSs, traffic participants are modeled as independent twin agents, containing virtual models replicating the physical objects and artificial intelligence for predictive decision-making. "Customization" addresses diverse and complex user needs, while "sustainability" focuses on the long-term evolution of FDT systems. Computational optimization is crucial for achieving customized and sustainable multi-agent maintenance, ensuring that FDT systems adapt to user needs and support sustainable development over time.

- **Achieving Lightweight and Redundant-Free Replication of Large-Scale Video Footage**: In DTs, videos uploaded by multiple terminals often contain significant redundancy, consuming excessive storage space and network bandwidth while increasing the risk of exposing sensitive pedestrian-vehicle information. Redundancy can hinder semantic segmentation, reducing the responsiveness and safety of pedestrian-vehicle interactions. Therefore, de-redundancy algorithms are essential for constructing DTs and reducing storage consumption. To obtain lightweight and redundant-free replication, techniques such as data deduplication algorithms, model pruning, knowledge distillation, video compression, and semantic communication can be employed, promoting effective management of both pedestrians and vehicles.
- **Balancing Standardized and Personalized Model Extraction for Pedestrian-Vehicle Agents**: Achieving a balance between standardization and personalization is crucial in pedestrian-vehicle interactions. Standardized agent models should support some industry standards such as UL-4600[1] and can capture general features,

[1] UL-4600 is available at https://www.intertek.com/automotive/ul-4600.

such as typical pedestrian behavior patterns and vehicle operation modes. In contrast, personalized agent models focus on individual differences, including walking patterns, driving styles, and vehicle configurations. Although commonalities exist, significant individual variations must be considered to model pedestrian-vehicle interactions precisely. Standardized models depend on extensive data collection, whereas personalized models require individual feature extraction via deep neural networks. An imbalance between the standardized and personalized DT model construction in ITSs may result in low-quality global and local models within FDT construction. This significantly degrades the accuracy of both (i) the global decision-making, e.g., traffic control and emergency management; and (ii) the local decision-making, e.g., customized navigation and travel planning.

- **Forward-Thinking and Spatially Versatile Cross-Model Migration under Ubiquitous Mobility**: Predicting user behaviors in DT models for highly mobile users, such as pedestrians and drivers, is challenging due to its inherent unpredictability [52]. For instance, drivers might deviate from suggested routes due to personal preferences or unforeseen circumstances, complicating the transfer of models. To tackle this issue, generative AI (GAI) technology can simulate potential user behaviors and decision-making paths, aiding in scenario forecasting [34]. Transfer learning techniques can also be employed to personalize and refine the model, enhancing its adaptability to individual user needs [24]. This approach facilitates effective management of interactions between pedestrians and vehicles, ultimately delivering more intelligent and tailored services.

5.3.2 System Implementation

The SV-FDT framework presented in Sect. 5.2 shows that cloud-edge-end collaboration is critical for real-time data integration and distributed processing in dynamic traffic environments, and continuous model refinement and data management can ensure local responsiveness and global coherence. In this section, we describe in detail the implementation of the proposed SV-FDT framework.

5.3.2.1 DT Construction by Cloud-Edge-End Collaboration

Figure 5.2 illustrates an implementation of the SV-FDT framework, which incorporates seven key steps: real-time video acquisition and integration, time-sensitive transmission, self-supervised semantic segmentation, pedestrian-vehicle interaction modeling, and local and global DT model construction. It also explicitly visualizes the data flow across functional modules at the end, edge, and cloud layers. SV-FDT leverages the CARLA simulation environment to create traffic scenes and develop local and global DT models. Prior systems primarily consist of real-time video acquisition, time-sensitive transmission, and DT model construction. SV-FDT addresses key gaps by incorporating self-supervised semantic segmentation, semantic-to-code

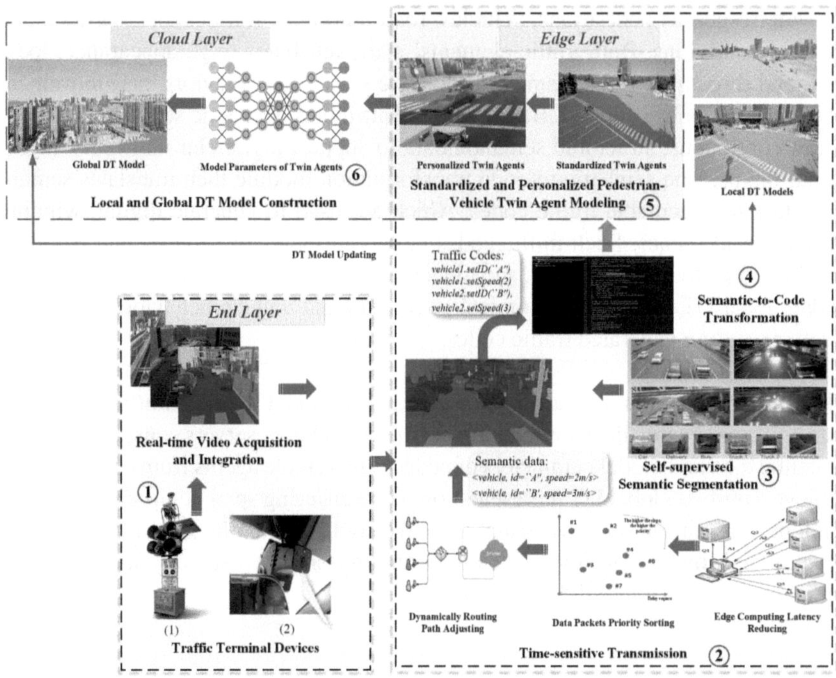

Fig. 5.2 Systematic framework of the proposed SV-FDT. The semantic segmentation module extracts vehicle data, e.g., *<vehicle, id="A", speed=2 m/s>*. The semantic-to-code transformation module converts this data into executable traffic codes: *vehicle1.setID("A"), vehicle1.setSpeed(2))*. These codes are subsequently processed and executed by both local and global DT models

transformation, pedestrian-vehicle interaction modeling, and real-time digital twin construction. CARLA, an open-source simulator built on Unreal Engine, generates high-fidelity traffic environments and accurately simulates traffic participant behaviors. SV-FDT utilizes CARLA's map editor to design immersive traffic scenarios featuring vehicles, pedestrians, and detailed surroundings.

The end layer is responsible for real-time video data acquisition via HD cameras and sensors, ensuring timely traffic perception. At the end layer, SV-FDT employs HD cameras and other terminal devices to collect and integrate real-time surveillance videos, establishing a foundation for constructing local and global DT models.

The edge layer handles critical processes, such as self-supervised semantic segmentation, semantic-to-code transformation, pedestrian-vehicle interaction modeling, and local DT model construction, enabling local-area traffic DT. At the edge layer, before semantic segmentation, edge nodes perform essential data preprocessing, including redundancy elimination, data integration, and synchronization. After that, semantic segmentation algorithms are applied to identify traffic elements in the video, such as pedestrians, vehicles, road markings, traffic signs, barriers, and weather conditions. Both traffic elements' static attributes (e.g., appearance, location, and direction) and dynamic attributes (e.g., speed, behaviors, and movement trajectories) can be captured. Based on these attributes, semantic segmentation generates continuous, structured semantic data to support a granular understanding of visual inputs. The semantic-to-code transformation module then translates semantic data into executable traffic codes, which are used to generate digital twinning of real-world scenes. Such traffic codes define attributes of vehicles and pedestrians, including locations, speeds, and movement directions. Edge nodes continuously update local DT models and simulate pedestrian and vehicle behaviors by successively executing generated traffic codes.

The cloud layer is a central hub for global DT model aggregation via semantic data and cross-region decision-making. At the cloud layer, servers constantly receive semantic data and model parameters of pedestrian-vehicle agents from edge nodes to achieve a unified global DT representation while ensuring spatiotemporal coherence without transmitting sensitive visual data. We could exploit the global DT model to develop and implement ITS applications running on cloud servers, such as traffic signal control and traffic flow prediction.

5.3.2.2 Implementation of Local and Global Digital Twins

Real-Time Video Acquisition and Integration: Video compression standards, such as H.264 and H.265, are applied to reduce bandwidth consumption while maintaining high video quality. Adaptive bitrate streaming adjusts video quality dynamically according to bandwidth availability, ensuring seamless streaming. With the time synchronization protocol, all video frames and sensor data are timestamped for accurate synchronization during analysis. Multi-sensor fusion algorithms, such as Kalman Filters, are used to integrate data from various sources to ensure synchronized object representation across modalities.

Time-Sensitive Transmission: URLLC technologies are employed to enable immersive interactions in DTs. Edge computing further minimizes latency by positioning computational resources closer to data sources (e.g., HD cameras and radars), enabling parallel processing of multi-modal data and reducing transmission delays. Based on latency requirements and importance, we prioritize data packets such as real-time perception, semantic understanding, and signal timing control, ensuring critical packets are transmitted first. We dynamically adjust transmission rates and route paths according to real-time network conditions and communication needs.

Self-supervised Semantic Segmentation: We prefer self-supervised semantic segmentation over unsupervised methods due to its ability to deliver real-time, pixel-level insights into traffic elements without manual labeling, which is time-consuming and costly. Advanced models such as DINO[2] excel at capturing long-range dependencies and contextual relationships, making them ideal for complex traffic scenarios. Specifically, we enhance semantic perception through a combination of contrastive learning for discriminative feature extraction, temporal correlation modeling for consistent object tracking, and context-aware refinement to suppress background interference. To mitigate the effects of occlusion and limited viewpoints, we implement cross-camera panoramic feature fusion and Transformer-based attention mechanisms that recover missing spatial information. Additionally, granularity-controllable segmentation allows precise customization of output detail to align with user requirements [55]. Techniques such as contrastive learning enhance the model's ability to identify key attributes at the pixel level [6]. For example, optical flow measures pedestrian and vehicle speeds, while pixel-level annotations precisely position traffic elements. By extracting static and dynamic attributes through semantic segmentation, we generate continuous, structured semantic data to support a detailed visual understanding of traffic scenes. This semantic data is then converted into traffic codes via the semantic-to-code transformation module.

Semantic-to-Code Transformation: The semantic-to-code transformation module is a core component of the framework, converting structured semantic data into executable traffic codes for CARLA simulation. This process uses rule-based algorithms and advanced deep learning techniques, e.g., generative AI and large language models (LLMs), to generate traffic codes based on traffic elements and their attributes, as illustrated in Fig. 5.2. The transformation logic ensures accurate scene replication within CARLA environments by adhering to CARLA's APIs and command structures. We continuously monitor traffic code performance in CARLA to maintain accuracy, refining the transformation logic as needed to correct discrepancies. This iterative process ensures that the traffic codes are both precise and reliable.

Pedestrian-Vehicle Interaction Modeling: In the CARLA environment, the vehicular motion model is based on a kinematic bicycle model, enabling realistic simulation of acceleration, deceleration, lane changes, and turning behaviors. To explicitly model pedestrian dynamics, a collision-free velocity model is employed [46], allowing pedestrians to exhibit natural walking patterns such as acceleration, deceleration, hesitation, and spontaneous crossings in response to nearby vehicles and infrastructure layouts. We develop pedestrian-vehicle twin agents with both standardized and personalized features to simulate realistic interactions. Using model-driven agent modeling, we treat traffic participants as independent, autonomous agents, combining standardized traits for predictable behaviors in typical scenarios with personalized features derived from individual-specific data. Each twin agent operates autonomously in the CARLA environment, independently perceiving its surroundings, making decisions, and executing actions. This hybrid approach balances standardized and personalized behaviors, enabling agents to dynamically adjust and produce realistic, responsive interactions in complex traffic environments. To enhance

[2] https://github.com/IDEA-Research/MaskDINO.

the system's ability to model highly unpredictable behaviors of pedestrians and vehicles, such as sudden pedestrian road crossings and abrupt stops, the agent models integrate two core technologies: semantic-level interaction representation and a probabilistic behavior inference module. The semantic-level interaction representation captures high-level intentions of both pedestrians and vehicles, including actions like waiting, crossing, and hesitating. To address inherent randomness and spontaneity, the probabilistic behavior inference module employs Gaussian process regression to generate behavioral predictions. A context-aware attention mechanism further refines these predictions in real time by adapting to interactions with surrounding agents. These components enable SV-FDT to effectively capture and respond to unpredictable behaviors, enhancing the system's realism and adaptability in complex urban environments.

Local and Global DT Model Construction: Local and global DT models are constructed and updated using semantic data and model parameters of twin agents. Edge nodes build 3D local panoramic DT models of roads, buildings, and infrastructure, either online or offline, and continuously update them by executing traffic codes. These traffic codes dynamically adjust attributes, such as weather, vehicle positions, and pedestrian presence, ensuring that local DT models remain accurate and up-to-date. The global DT model aggregates semantic data and agent parameters from distributed edge nodes, providing a unified, large-scale view of the transportation network. Data exchanges between edge nodes and cloud servers are designed to maintain privacy and spatiotemporal coherence by omitting sensitive visual information. As local DT models update in real time, the global DT model synchronizes these updates to preserve a consistent and coherent representation of traffic conditions across regions. Local DT models mirror specific traffic regions in ITS applications, such as navigation guidance, by incorporating pedestrian-vehicle interactions and evolutionary reasoning based on real-time updates. The global DT integrates these updates to provide a comprehensive network view for guiding vehicles and pedestrians along the most efficient or safest routes.

To meet the real-time demands of ITS applications, we introduce a latency-aware synchronization mechanism to coordinate local and global DT models. The total update delay, T_{update}, includes transmission, processing, and aggregation delays. To ensure timely responsiveness, we impose the constraint:

$$T_{\text{update}} \leq T_{\text{threshold}} \tag{5.1}$$

where $T_{\text{threshold}}$ is empirically set to 0.5 s based on safety-critical latency requirements in ITS environments. To satisfy this constraint, SV-FDT adopts URLLC protocols and packet prioritization techniques to reduce transmission delays. Time-sensitive packet scheduling dynamically prioritizes critical semantic data and traffic control messages to ensure they are transmitted first. This mechanism enables SV-FDT to maintain reliable synchronization and real-time DT updates across system layers.

5.3.3 *Case Study and Experimental Results*

Given the overall architecture of the proposed SV-FDT framework, this section presents a pioneering case study on optimizing traffic light time setting, addressing design requirements and challenges as presented in Sect. 5.2.

5.3.3.1 Experimental Setup and System Design

As illustrated in Fig. 5.3, we developed a testbed platform based on SV-FDT to optimize traffic signal timing in ITSs. The end layer includes HD cameras, radars, and traffic lights. The cameras and radars collect visual data, which they preprocess and integrate locally. The edge layer comprises a desktop computer serving as the traffic signal control center and five laptops equipped with NVIDIA RTX 4060 GPUs and i9-14900HX processors, functioning as edge nodes. To enable flexible deployment of edge nodes under resource-constrained conditions, we employ lightweight semantic segmentation algorithms and large language models with significantly reduced parameter sizes based on available edge resources. One edge node runs a lightweight semantic segmentation algorithm to monitor pedestrian and vehicle positions, generating real-time semantic data on their speeds and trajectories. A second node handles semantic-to-code transformation, converting this data into executable traffic control commands for the CARLA simulation environment. A third node is responsible for 3D scene modeling, generating a panoramic representation of the road and surrounding structures. The fourth and fifth edge nodes maintain local DT models for two subregions, updating them in real time. The traffic signal control center uses these local DT models to optimize signal settings dynamically.

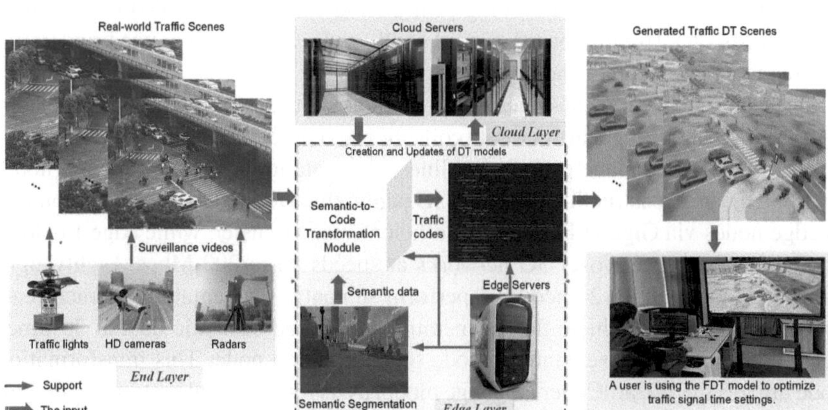

Fig. 5.3 The SV-FDT framework-based FDT testbed platform for optimizing traffic signal timing

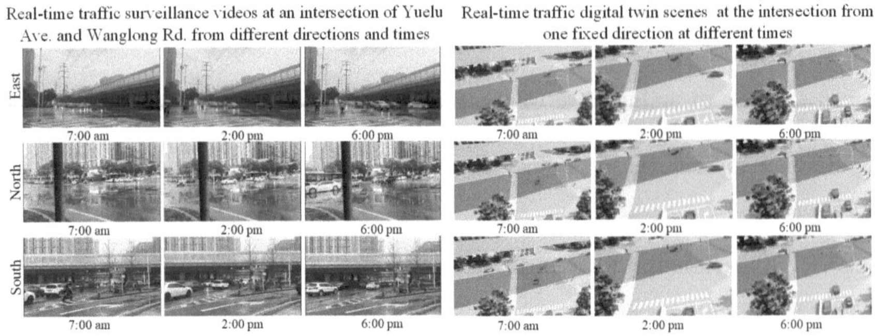

Fig. 5.4 Real-time traffic digital twin scenes at the intersection of Yuelu Ave. and Wanglong Rd. at three different times of day (Time: 7:00 AM, Morning Peak Hour; Time: 2:00 PM, Midday Off-Peak; Time: 6:00 PM, Evening Peak Hour), generated by the proposed SV-FDT model based on real-time traffic surveillance videos from three cameras

The traffic signal control center employs a fuzzy clustering algorithm combined with a density-based spatial clustering algorithm (FCA-DBSCAN) [1] to estimate pedestrian and vehicle densities. FCA-DBSCAN leverages traffic structure, real-time pedestrian and vehicle densities, road width, and pedestrian walking speed to enable adaptive signal control. The cloud layer hosts two servers: one for generating the global DT model and the other for executing global traffic signal optimization. Figure 5.4 illustrates the deployment of the SV-FDT system at the intersection of Yuelu Ave. and Wanglong Rd. in Yuelu District, Changsha, China. Real-time federated digital twin scenes are generated from traffic surveillance videos captured by three cameras. Due to substantial traffic fluctuations throughout the day, this location serves as an ideal urban testing site. To evaluate SV-FDT's dynamic signal control capabilities, experiments were conducted at 07:00, 14:00, and 18:00, representing the morning peak, off-peak, and evening peak periods, respectively. SV-FDT dynamically adjusts signal timings based on road width, pedestrian walking speed, and traffic volume to enhance safety and optimize traffic flow.

The experiments utilized 4-megapixel cameras capturing traffic with a bandwidth of 10 Mbps, supporting real-time data collection and processing at 30 frames per second. Drone aerial footage provided offline video data to the scene modeling node, enabling the pre-construction of a static 3D panoramic DT model. End nodes connect to edge nodes via Gigabit Ethernet for efficient video transfer, while edge-to-cloud communication occurs over 5G networks at speeds up to 200 Mbps. Lightweight UniMatch V2+ DINOv2 encoders[3] perform semantic segmentation to track real-time pedestrian and vehicle flow, generating structured semantic data at 30 times per second sent to the semantic-to-code transformation node. This transformation node utilizes the WizardCoder language model as its foundational pretraining model. After fine-tuning with semantic data and traffic codes, WizardCoder generates executable traffic codes from structured semantic inputs. The local DT node loads the

[3] https://github.com/LiheYoung/UniMatch-V2.

3D panoramic model from the scene modeling node and uses CARLA to execute these codes, simulating real-world pedestrian-vehicle interactions at the frequency of 30. Meanwhile, the global DT modeling node integrates semantic data from all local DT models, creating a comprehensive global DT model within CARLA. The aggregation and updates of the global DT are done less frequently than those of the local DTs. Instead, updates occur periodically every minute or when significant changes in traffic patterns, such as traffic events, are detected. Ultimately, SV-FDT dynamically adjusts signal timings to improve road safety. The case study showcases the feasibility and validity of the SV-FDT framework.

5.3.3.2 Performance Evaluation

Performance evaluation of the proposed platform is based on seven key metrics: (i) **Mirroring Delay**: the time required to replicate real-world traffic conditions within the digital twin system; (ii) **Recognition Accuracy**: the proportion of correctly identified vehicles and pedestrians in surveillance video frames; (iii) **Traffic Flow under Different Signal Modes**: the volume of vehicles passing through the intersection under various traffic signal control strategies; (iv) **Mirroring Delay Intervals under Varying Confidence Levels**: the distribution of mirroring delays at different confidence thresholds; (v) **Delay under Different Resource Constraints**: the latency performance under limited computational or bandwidth resources; (vi) **Signal Timing Settings under Varying Recognition Accuracy Ratios**: traffic signal duration settings corresponding to different recognition accuracy ratios; (vii) **Subjective Evaluation**: based on user feedback regarding video fluency, cyber-physical consistency, and movement synchronization. Based on experimental data at the intersection of Yuelu Ave. and Wanglong Rd., Fig. 5.5 compares SV-FDT with the traditional terminal-server framework across the above performance metrics.

As shown in Fig. 5.5a, the SV-FDT framework outperforms the traditional terminal-server framework in DT modeling, achieving significantly lower average delay (AVE-DL), maximum delay (MAX-DL), minimum delay (MIN-DL), and jitter (JIT-DL). This improvement is attributed to two key reasons in the proposed SV-FDT: (i) data processing occurs at edge nodes closer to the data source; and (ii) only traffic codes and model parameters of twin agents, rather than raw surveillance videos, are transmitted to the cloud. Figure 5.5b shows the recognition accuracy comparison, where SV-FDT also performs better, demonstrating higher average (AVE-RA), maximum (MAX-RA), and minimum recognition accuracy (MIN-RA), along with lower jitter (JIT-RA). This is because the semantic segmentation algorithm in SV-FDT is executed at edge nodes, which can continuously learn from local traffic environments, ensuring overall recognition accuracy. Figure 5.5c illustrates traffic flow variations under fixed and adaptive signal modes at different times of the day. Under the adaptive signal mode, the average vehicle and pedestrian densities during peak hours are 1.8× and 10× higher than those during off-peak periods, and 1× and 10× higher than those during morning peak hours, respectively. Compared to the fixed signal mode, the adaptive mode effectively enhances traffic flow by dynamically adjust-

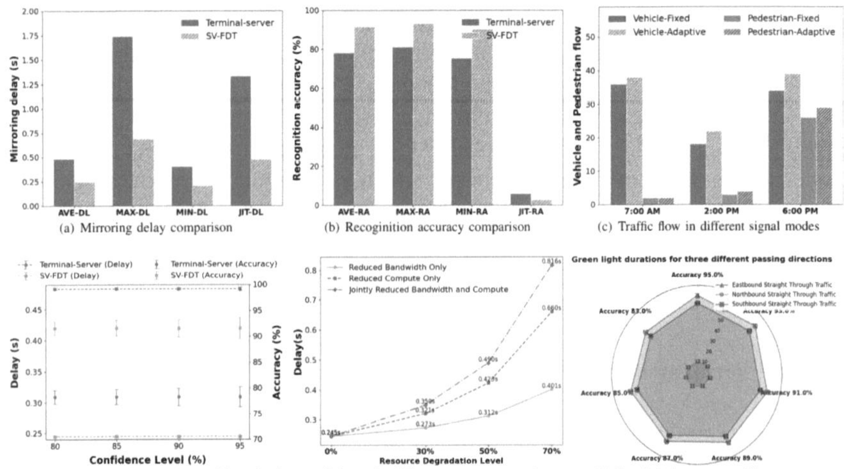

(a) Mirroring delay comparison (b) Recoginition accuracy comparison (c) Traffic flow in different signal modes

(d) Delay intervals under varying confidence levels (e) Delay under different resource constraints (f) Signal timing vs. recognition accuracy

Fig. 5.5 Comparisons on objective measurements

Table 5.1 Comparisons on subjective measurements

Framework	Video fluency	DT consistence	Sync.
Terminal-server	Choppy	Discrepant	Asynchronous
SV-FDT	Fluent	Consistent	Synchronous

ing the signal timings, significantly increasing vehicle and pedestrian throughput during peak hours. Figure 5.5d presents the mirroring delay intervals at different confidence levels. At confidence levels above 80%, SV-FDT achieves a mirroring delay of approximately 0.24 s while maintaining recognition accuracy within the range of [0.90, 0.93]. Figure 5.5e illustrates mirroring delay variations under three different resource constraint modes. The results show that the total mirroring delay remains below 0.5 s even when computational or bandwidth resources are degraded to 50%, individually or jointly. Figure 5.5f shows the relationship between signal timing settings and recognition accuracy ratios. For simplicity, green light durations assigned to three different passing directions are selected as illustration samples. When recognition accuracy ratios are higher, signal timing configurations remain almost unchanged. In summary, the SV-FDT model outperforms the terminal-server framework in both delay and recognition accuracy, while demonstrating high robustness and stability. Compared with fixed traffic signal settings, SV-FDT achieves significant performance improvements by adopting adaptive traffic signal timing. Finally, subjective evaluations were conducted by 10 volunteers on the test platform. Table 5.1 shows that the terminal-server framework encountered issues related to DT modeling delay, inconsistencies, and asynchronous movement, whereas the SV-FDT framework operated seamlessly. These findings confirm that our framework provides an outstanding and immersive quality of experience (QoE) in both objective and subjective measurements.

- *High-Precision DT Extraction via Multi-modal Fusion*: Extracting and fusing features from multi-modal traffic data (e.g., images, audio, text, videos) enhances the accuracy of FDT models in heterogeneous ITS environments. While traditional methods rely on single-modal data, multi-modal models require real-time data acquisition and distributed storage, posing privacy and security challenges. Future work will focus on building FDT models using knowledge graphs and deep learning to improve security and accuracy.
- *Seamless DT Synchronization Across Diverse Vehicular Communication Systems*: Vehicular communication devices vary in transmission rates, coverage, reliability, and interference. Real-time DT synchronization across these vehicular communication systems is challenging. Future efforts will leverage 6G technology for efficient wireless transmission and explore visible light communication (VLC), radio frequency identification (RFID), millimeter wave (mmWave), acoustic communication, and low Earth orbit (LEO) satellites to optimize DT synchronization.
- *Extending FDT to Future ITSs with Hyper-Spatial Mobility*: As AI advances, future ITSs will evolve into hyper-spatial systems supporting efficient and secure transportation across rail, subway, drone, and autonomous vehicle networks. Future research will investigate deep learning and intelligent control technologies to develop hyper-spatial FDT models for autonomous decision-making, route planning, state monitoring, performance evaluation, and fault prediction across multiple transport modes. In addition, we will explore integrating resource scheduling strategies, such as TD3 [33], into these systems to optimize computational resource allocation, improve data processing and communication efficiency, and ensure real-time performance with reduced latency in large-scale, multi-source environments.
- *Optimal Traffic Signal Control in Self-evolving FDT Systems*: Errors in DT models can significantly impair traffic signal accuracy and decision-making, making optimal FDT coordination highly complex. Future research will focus on developing mechanisms to detect and correct inaccuracies in traffic DT systems, applying filtering techniques, and leveraging federated learning and collaborative optimization to enhance data sharing and coordinated decision-making. For instance, in multi-intersection signal control, local SV-FDT agents will use reinforcement learning to adjust signal timings based on real-time pedestrian-vehicle interactions [29]. Instead of transmitting raw data, only model updates are shared via federated learning, enabling the cloud to construct a global control policy. Distributed, multi-agent learning strategies will be also developed to adaptively respond to real-time traffic, improving both traffic flow and road safety. Additionally, we will explore adaptive learning mechanisms that allow SV-FDT to dynamically respond to diverse traffic scenarios. In future work, SV-FDT will be equipped with reinforcement learning capabilities and semantic feedback loops to enable real-time policy updates based on local observations of traffic patterns, weather conditions, pedestrian flow, and vehicle behaviors. These localized models will be aggregated via federated learning to refine global optimization strategies, allowing SV-FDT to adapt to varying conditions across different regions while preserving data privacy.

5.4 AIGC-Driven Real-Time Interactive 4D Traffic Scene Generation in Vehicular Networks

5.4.1 System Model and Design Goals

5.4.1.1 System Architecture

The proposed AIGC-driven 4D traffic scene generation system for vehicular networks consists of three layers: the terminal layer, the edge layer, and the cloud layer, as shown in Fig. 5.6. The terminal layer focuses on collecting real-time data from sensors and vehicles; the edge layer processes this data locally to ensure quick response and handles latency-sensitive tasks; and the cloud layer manages resource-intensive tasks, such as model pretraining and large-scale data storage, providing refined model updates to the edge for real-time adaptability. The framework encompasses five key computational processes: data collection, management, pretraining, fine-tuning, and inference, which are described as follows. Unlike existing architectures, this framework leverages a cloud-edge-terminal collaboration and text-to-code models to ensure real-time interactivity, temporal coherence, and adaptability, addressing the dynamic demands of intelligent transportation systems.

(1) **Data Collection**: Data collection is the first step for enabling real-time, interactive 4D traffic scene generation within an AIGC-driven system. This process,

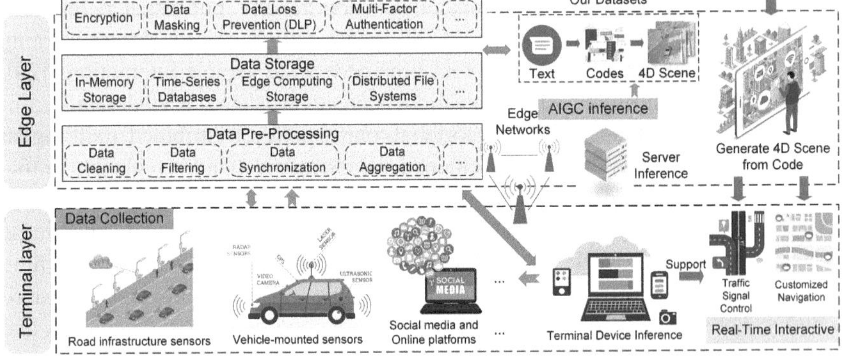

Fig. 5.6 The proposed vehicular network framework for AIGC-driven real-time interactive 4D traffic scene generation system consists of the terminal layer, edge layer, and cloud layer. All these support the 4D traffic scene generation with AIGC lifecycle, i.e., data collection, management, pre-training, fine-tuning, and inference

primarily handled by terminal devices, involves gathering data comprehensively from various sources. The quality and scope of the collected data are critical for generating accurate and responsive traffic scenes. Key data sources include real-time traffic information from road infrastructure sensors, and GPS systems, as well as environmental data from meteorological stations and road surface sensors, vehicle-mounted sensors and smart devices, which are vital for capturing the dynamic nature of transportation systems. Additionally, integrating non-traditional data sources, such as social media and online platforms [50], provides valuable insights into human behavior and urban activities, thereby enhancing the realism and interactivity of the generated DT scenes [35, 47]. By combining these diverse data collection strategies, the system ensures that AIGC-driven 4D traffic scenes are accurate and capable of adapting in real-time to evolving conditions.

(2) **Data Management**: Effective data management is crucial for enabling efficient information sharing and integration among stakeholders while protecting sensitive information throughout its lifecycle. This process involves handling diverse data types, including real-time traffic information, simulated data, AIGC-created information, and historical traffic records. It begins with comprehensive data preprocessing, which includes real-time cleaning, filtering, synchronization, spatial processing, aggregation, feature extraction, and image and video processing. These steps ensure the data are accurate, consistent, and ready for immediate use. The preprocessed data is then stored using high-performance technologies, such as in-memory storage, time-series databases, edge computing storage, and distributed file systems. This robust storage infrastructure supports the continuous training and updating of AIGC models, enabling the seamless generation of real-time, interactive 4D traffic scenes. Given the data sensitivity involved, strict measures are necessary to prevent privacy violations, unauthorized surveillance, and public security risks.

(3) **AIGC Pre-training**: Selecting suitable AIGC models is essential for achieving real-time, interactive 4D traffic scene generation. Text-to-code models are favored over text-to-video models because they produce adaptable, executable codes that dynamically responds to real-time data, making them ideal for complex, evolving traffic environments. These models are computationally efficient and integrate seamlessly with existing ITSs, enhancing overall system responsiveness. Pre-training for these models usually occurs on cloud servers with significant computational resources. Advanced pretraining techniques can further optimize text-to-code models for real-time 4D traffic scene generation. For instance, transformer models capture the sequential and temporal dynamics of traffic data, enabling the generation of contextually accurate and responsive codes [42]. Additionally, code-specific embeddings refer to representations of code (such as source code), which ensure that the generated codes are contextually appropriate by accurately representing the syntax and semantics of the relevant programming languages. Large-scale pretraining on diverse code corpora equips the model with the versatility to handle various programming challenges in dynamic, real-time interactive traffic management systems.

(4) **AIGC Fine-Tuning**: AIGC fine-tuning customizes pre-trained models for specific tasks. Using curated datasets, fine-tuning can occur either in cloud environments or at the edge. Advanced fine-tuning techniques are employed to optimize text-to-code models for real-time, interactive 4D scene generation, allowing for dynamic, user-driven creation and modification. Given the ever-changing nature of transportation scenes, which involve moving vehicles and shifting environments, techniques like multi-modal learning and alignment are essential for converting text inputs into temporally coherent sequences that accurately represent object trajectories and interactions. Instruction tuning is critical for text-to-code models because it allows the model to understand better how users might describe their coding needs. Additionally, data augmentation can expand the variety of natural language descriptions and code examples available for fine-tuning. It is essential when a large amount of task-specific training data is impossible. Fine-tuning for scene composition enables the model to generate modular components (e.g., roads, buildings, vehicles) that can be assembled and modified in real time based on text inputs.

(5) **AIGC Inference**: The inference phase utilizes pre-trained and fine-tuned AIGC models to generate content from specific inputs while minimizing computational overhead efficiently. This optimization enables AIGC inference to be deployed on edge servers or devices, such as in-vehicle systems and roadside units, facilitating real-time interactive services. The content generated is essential for creating real-time interactive 4D transportation scenes, which support comprehensive analysis and predictive modeling for effective traffic management and optimization. By deploying these models directly within the transportation infrastructure at the edge, the system can dynamically generate and update 4D scenes in real time, accurately reflecting current traffic conditions and environmental changes. This approach significantly enhances the responsiveness and precision of traffic management systems, supporting real-time interactive ITS applications.

5.4.1.2 Example Scenarios

AIGC-Driven Real-Time Interactive 4D Traffic Scene Generation Empowers Customized Navigation: Customized navigation is a critical application in intelligent transportation systems. AIGC-driven navigation systems address significant challenges, showcasing their potential in customized navigation. For example, Kuribayashi et al. [21] proposed a vision-and-language model that uses LLMs to interpret user instructions and generate code for robotic navigation. However, these AIGC methods cannot effectively replicate the complex interactions between driving vehicles and surrounding environments, resulting in poor navigation experience.

AIGC-driven real-time interactive 4D scene generation offers high-fidelity navigation by continuously updating driving scenes in real time. As shown in Fig. 5.7, the AIGC-driven system enables vehicle drivers to visualize driving paths, traffic participants, and surrounding environments. The system enhances safety and efficiency by offering a personalized, immersive navigation experience, meanwhile improv-

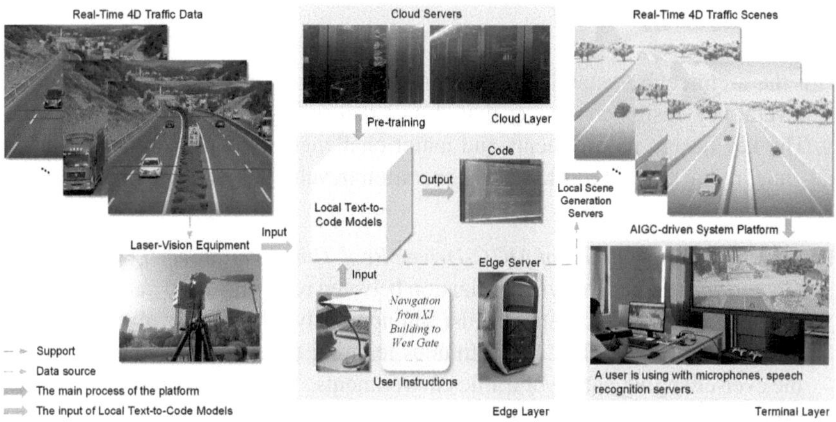

Fig. 5.7 Practical implementation of the AIGC-driven real-time interactive 4D traffic scene generation framework

ing the accuracy and dependability of customized navigation while expanding their applicability to diverse, complex environments. As a result, AIGC-driven real-time interactive 4D traffic scene generation systems represent a significant advancement in intelligent transportation, providing vehicle drivers with adaptive guidance tailored to their needs and ensuring real-time responsiveness and interactivity.

AIGC-Driven Real-Time Interactive 4D Traffic Scene Generation Assists Smart Transportation Management: Smart transportation management optimizes traffic signal control timings, reduces congestion, and enhances overall traffic flow efficiency. AIGC captures vehicle and pedestrian movements, enabling dynamic real-time adjustments to traffic signal timings. A prominent example is the cross-city meta multi-agent traffic signal control model [18]. This model uses a dual-layer Transformer architecture to improve adaptability across urban environments. Wei et al. [44] introduced a generative adversarial network to enable more accurate signal adjustments. However, these AIGC methods lack the ability to simulate complex traffic scenarios, and operating in solos limits their effectiveness in adjusting traffic signals.

AIGC-driven real-time interactive 4D scene generation systems digitally twin vehicle movements, traffic light sequences, and environmental conditions in real time. Thus, the AIGC-driven system improves traffic signal timing, congestion management, and efficiency. As shown in Fig. 5.7, traffic police can adjust traffic signal controls within the system, which autonomously generates corresponding 4D scene simulations. These simulations allow detailed analysis of traffic congestion and pedestrian-vehicle interactions, enabling dynamic optimization of signal settings. Integrating real-time interactive 4D scenes allows cities to adopt more intelligent, responsive systems, paving the way for innovative urban transportation management.

5.4.1.3 Key Design Goals

With the architecture established, implementing this framework requires addressing several key design challenges that impact real-time performance and interactivity. The key design requirements and major challenges of deploying AIGC-driven real-time interactive 4D traffic scene generation in vehicular networks are discussed below.

(1) *Adaptive Self-evolution of AIGC-Driven Real-Time Interactive 4D Traffic Scene Generation*: To ensure synchronization between real-world and generated traffic scenes, AIGC-driven systems must autonomously adapt and evolve. This adaptive self-evolution relies on continuous learning from real-time data, reflecting the ever-changing nature of traffic environments. The system must continuously process and integrate data from various sources to stay aligned with current traffic conditions, environmental changes, and emerging trends. Our framework addresses this need through a cloud-edge-terminal collaborative structure, which optimally distributes processing tasks. By dynamically updating its models in response to new training data, the AIGC-driven system ensures that the generated 4D traffic scenes accurately reflect the current state of the transportation network. Additionally, this capability allows the system to recognize and adapt to long-term trends, ensuring its predictive algorithms remain effective as traffic patterns evolve. The self-evolution process must be scalable and flexible, allowing the system to seamlessly transition across different geographic regions and respond effectively to diverse traffic conditions. Key aspects of this process are highlighted below.

(2) *Generating High-Precision and Fast-Responsive Contextual Text-to-Code*: Text-to-code models must produce executable code that captures the complexity and dynamics of traffic scenes while adapting to environmental changes in real time. It requires a pretraining process that thoroughly understands traffic contexts, captures nuances in traffic patterns, and generates code that seamlessly adapts to new inputs and evolving conditions. The challenge is ensuring the model consistently generates contextually relevant, efficient, and adaptable code for real-time execution, while preserving the fidelity and accuracy of the 4D traffic scenes. Enhancing pretraining with diverse traffic datasets can improve the model's ability to generate contextually appropriate code. Additionally, incorporating a reward mechanism can encourage the model to produce code that meets real-time execution requirements while maintaining high fidelity and accuracy.

Designing Efficient Data Acquisition Schemes for Dynamic Fine-Tuning: Dynamic fine-tuning enables models to continuously learn and adapt to new traffic data, evolving conditions, and specific domain requirements. Efficient data acquisition is essential for maintaining temporal and spatial coherence in 4D scene generation, ensuring outputs align with domain-specific needs, user demands, and application contexts. Aligning fine-tuning with transfer learning enhances this coherence. Techniques like model distillation and pruning optimize performance in resource-constrained environments, such as vehicular networks, without compromising efficiency. However, the availability of high-quality and diverse data remains a significant challenge. Limited access to diverse traffic data impairs the model's ability to generalize across different scenarios, while low-quality or noisy data can lead to inaccurate code and reduce the effectiveness of fine-tuning. To address these issues, robust preprocessing methods such as anomaly detection and data normalization are employed to reduce noise and inconsistencies. Furthermore, a hierarchical data fusion approach integrating heterogeneous data sources like vehicle sensors, infrastructure devices, and weather feeds, can ensure a reliable foundation. Another critical challenge is managing the latency introduced by dynamic fine-tuning. In real-time 4D traffic scene generation, even slight delays in processing can result in outdated or incorrect outputs, severely impacting traffic management and safety. Edge computing is leveraged for localized fine-tuning tasks, and predictive algorithms is further adopted to address data quality issues proactively, enhancing system resilience.

Optimizing Networking Resources for Real-Time 4D Traffic Scene Generation: Real-time 4D traffic scene generation over vehicular networks demands high computational efficiency and minimal latency. Vehicular edge computing (VEC) enhances system performance by offloading computational tasks to roadside units (RSUs), minimizing reliance on cloud-based processing, and reducing communication latency [28]. In the context of 6G heterogeneous vehicular networks, edge computing enhances the efficiency of computation-intensive services by offloading tasks to local resources, reducing both latency and the burden on centralized cloud systems [16]. Dynamic edge offloading distributes workloads between the edge and cloud, enhancing system resilience [5, 22, 49]. However, managing resources in heterogeneous environments with varying computational capacities is challenging. Optimization strategies include using neural architecture search to deploy lightweight AIGC models tailored to edge-specific constraints. Techniques such as pruning and quantization reduce model size while preserving accuracy, enabling real-time inference on resource-limited devices. AI-driven resource scheduling dynamically allocates computational power based on traffic complexity and predictive analytics, further improving performance. These approaches ensure that 4D traffic scene generation remains responsive and scalable within dynamic vehicular networks.

(3) *Personalized Interactions in AIGC-driven Real-time Interactive 4D Traffic Scene Generation*: In real-world traffic, each participant (e.g., a vehicle, pedestrian, or cyclist) exhibits unique characteristics, preferences, and decision-making processes. By treating these participants as independent agents and personalizing their interactions, the system can more accurately capture the complexity of traffic dynamics. Personalized interactions enable the system to adapt to individual driving behaviors, such as route preferences, driving habits, and the way the driver interacts with other vehicles, leading to more precise predictions of traffic flow, congestion, and potential hazards. This personalization also accounts for external factors such as weather and road conditions, further refining driver

behavior models. Ultimately, personalized interactions make AIGC-driven 4D traffic scene generation systems realistic and applicable to real-world traffic scenarios.

Producing Tailored Interaction in Uncertain and Diverse Traffic Conditions: In real-world traffic, unpredictable conditions, such as dense fog on specific road sections, require the 4D simulation system to dynamically adapt interactions among traffic participants. Tailoring these interactions to environmental factors enhances realism and improves the system's ability to predict and manage risks. For instance, when vehicles enter or exit foggy areas, to account for reduced visibility, the system must adjust vehicle operations, such as increasing safe following distances and lowering driving speeds. The system must continuously fine-tune these interactions to maintain realism, generate accurate traffic flow predictions, and identify potential hazards. Any deviation from realistic interaction behaviors can compromise the simulation's accuracy, particularly in real-time scenarios. The simulation can deliver more effective traffic management and risk assessment by seamlessly adapting to these uncertain and diverse conditions.

Synthesizing Customized Interaction for Various Traffic Participants: Customized 4D traffic scene generation is essential for developing realistic traffic simulations. An AIGC-driven system personalizes traffic scenes to match each participant's characteristics and objectives, ensuring that the scenarios accurately replicate diverse real-world behaviors. To achieve this, the system must gather and integrate detailed data on individual users, such as driving habits, preferred routes, speed preferences, and typical responses to traffic signals and road conditions. Furthermore, it should model human driver behaviors precisely, accounting for personalized characteristics, reaction times, and responses to external factors like weather and road conditions.

Guaranteeing Immersive Quality of Experience in High-Hidelity Interactions: Ensuring immersive Quality of Experience (QoE) is a key focus in high-fidelity interactions in complex systems like 4D traffic scene simulations. Immersive QoE ensures that users perceive these simulations as realistic and engaging. Achieving this requires evaluation metrics that address both subjective and objective aspects of the user experience. Subjective metrics assess qualitative elements, such as the smoothness of movements and overall realism, while objective metrics measure quantifiable factors like traffic flow consistency and response times. To maintain high-fidelity interactions, the system must integrate these metrics and make real-time adjustments, such as smoothing vehicle trajectories or enhancing responsiveness to sudden environmental changes. By continuously optimizing based on these metrics, the system can deliver a more engaging and realistic user experience.

5.4.2 System Implementation

Based on the overall architecture of two applications in Sect. 5.4, this section presents the practical implementation of a AIGC-driven real-time interactive 4D traffic scene generation framework, addressing Sect. 5.3 's design requirements and challenges.

Figure 5.8 illustrates the practical implementation of the proposed real-time interactive AIGC-driven 4D traffic scene generation framework, which employs a cloud-edge-terminal collaborative architecture. This system integrates several key components: laser-vision equipment, microphones, speech recognition servers, local text-to-code models, local scene generation servers, and cloud servers.

- **Laser-Vision Equipment**: Tracks the real-time positions of pedestrians and vehicles.
- **Microphones**: Facilitate interaction by receiving spoken commands, such as navigation requests or weather updates.
- **Speech Recognition Servers**: Accurately convert spoken instructions into text, ensuring a seamless user experience.
- **Local Text-to-Code Models**: Deployed on edge servers equipped with NVIDIA A100 GPUs, these models handle AIGC fine-tuning and inference. They process

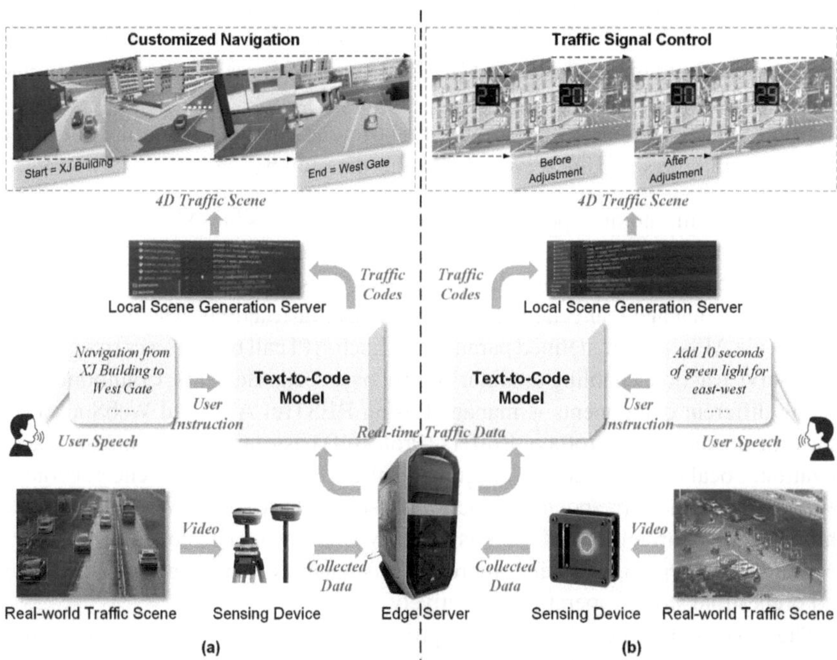

Fig. 5.8 AIGC-driven real-time interactive 4D traffic scene generation for customized navigation and traffic signal control applications, with the detailed data processing workflow

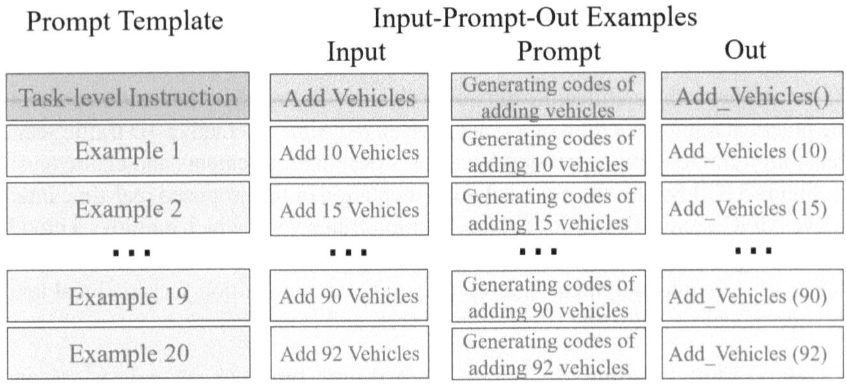

Fig. 5.9 User instruction and corresponding input-prompt-output pairs for fine-tuning training

user instructions and real-time traffic data from the speech recognition servers and laser-vision equipment to generate corresponding traffic codes.

- **Local Scene Generation Servers**: Running CARLA, an open-source simulator powered by Unreal Engine assets, these servers use the traffic code generated by the text-to-code models to create 4D traffic scenes. With the CARLA Map Editor, local scene generation servers digitally replicate real-world environments, creating immersive scenarios with vehicles, pedestrians, and detailed surroundings. The system continuously updates the CARLA model with real-time traffic data and user interactions.
- **Cloud Servers**: Manage pretraining of text-to-code models and transmit trained model parameters to edge servers for further fine-tuning.

In the system, atomic operations are basic and indivisible API functions, e.g., adding a number of vehicles and pedestrians, changing the weather from cloudy to sunny. These operations are modeled as microservices, each handling a specific task to efficiently manage real-time traffic data and user instructions. The microservices interact via APIs with predefined parameters. Each API call triggers a corresponding microservice action, enabling modular control over scene elements. Communication among different components is managed using RESTful APIs and WebSocket for real-time updates, with Protocol Buffers (ProtoBuf) ensuring efficient message serialization. Local text-to-code models generate the API codes, while scene generation servers execute the corresponding functions based on these codes[4] WizardCoder was chosen as the pretraining model, as its extensive training can be conducted on large programming corpora, including microservices, enables it to understand, generate, and orchestrate atomic operations effectively.

The fine-tuning process customizes pre-trained models for real-time 4D traffic scene generation, specifically targeting microservice orchestration. This fine-tuning

[4] The code is available at https://github.com/OpenHUTB/4D_SCENE_GEN.

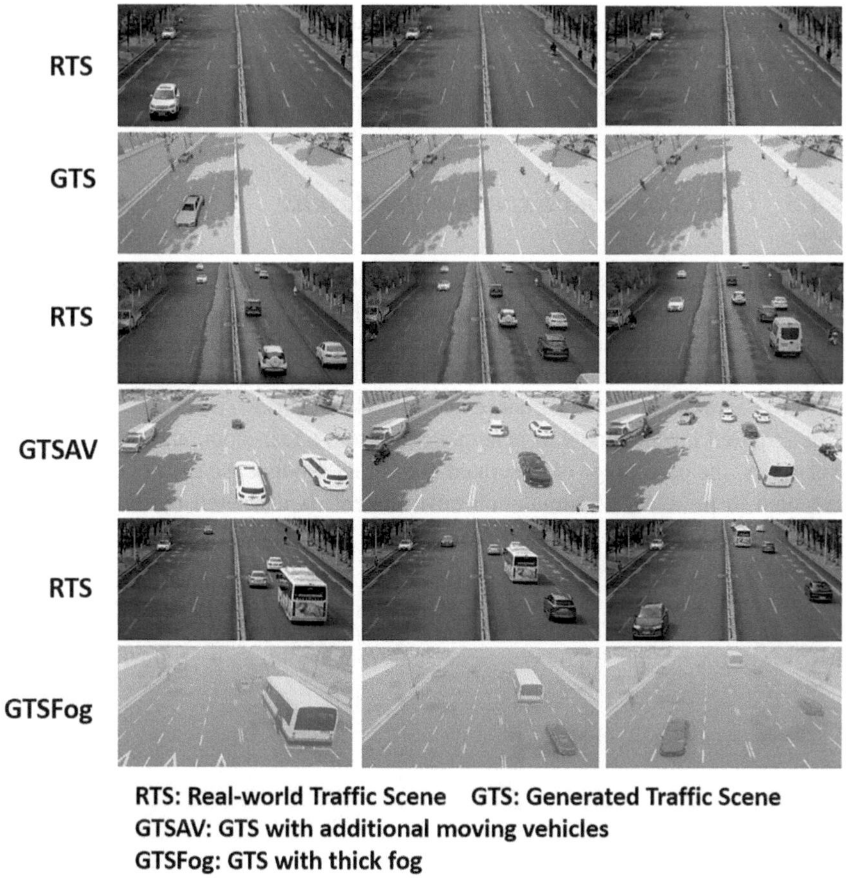

Fig. 5.10 Real-time interactive 4D traffic scenes generated by the AIGC-driven testbed system

uses specialized datasets that represent the system's microservices. Standard prompts are employed to build the fine-tuning datasets. We offer a microservice-based training dataset[5] for instruction tuning, including user instructions, prompts, and corresponding outputs. Every user instruction executable via an API call generates a minimum of 20 training samples comprising input-prompt-output pairs. As illustrated in Fig. 5.9, user instructions guide the system in executing specific tasks. After fine-tuning with vehicle-adding instructions, the model translates commands like "adding j vehicles" into the corresponding code "Add_Vehicles(j)". This atomic operation ensures the accurate execution of simple tasks. Additional training samples ensure precise code generation for complex instructions involving multiple API calls. Figure 5.10 illustrates the AIGC-driven testbed system, which generates real-time interactive 4D traffic scenes. The system accomplishes this by continuously monitoring and detecting

[5] The dataset is available at https://github.com/OpenHUTB/4D_SCENE_GEN/training_data.

traffic flow through laser-vision equipment. User interactions further enable tasks like adding vehicles to the road or changing weather conditions, highlighting the system's interactive capabilities.

5.4.3 *Case Study and Experimental Results*

To validate the proposed system, we evaluated its performance using two fine-tuned text-to-code frameworks: (i) a traditional framework that directly generates implementation code based on user instructions; and (ii) the proposed microservice-based framework, which generates and orchestrates API code from user instructions. We assessed their performance using two key metrics: (i) **Inference Time**, the time required for the model to generate code from a text input; and (ii) **Execution Accuracy**, the percentage of generated codes that produce the correct output when executed in the CARLA simulation environment, is measured by the percentage

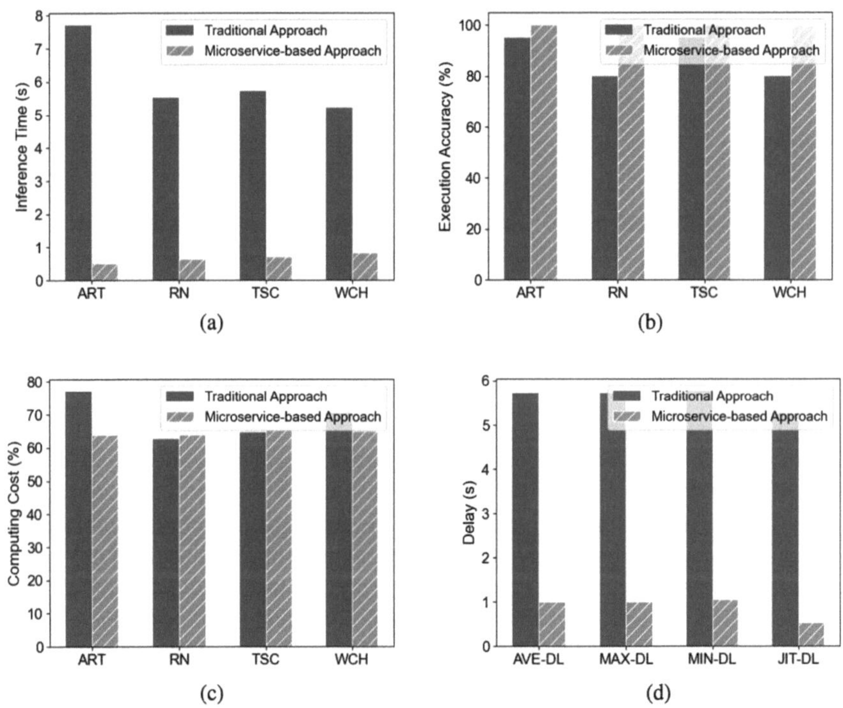

Fig. 5.11 Performance comparison of traditional versus microservice-based frameworks: **a** inferencing time, **b** execution accuracy, **c** computing cost and **d** Delay. ART = adding/removing traffic participants, RN = route navigation, TSC = traffic signal control, WC = weather changes. AVE-DL = lower average delay, MAX-DL = maximum delay, MIN-DL = minimum delay, and JIT-DL = jitter

of correct outputs. Inference Time measures system responsiveness, while Execution Accuracy ensures scene reliability; (iii) **Computing Cost**: the percentage of resource utilization; (iv) **Delay**: the time required to generate 4D traffic scene from speech. We constructed 80 user instructions for evaluation, covering tasks including adding/removing traffic participants, route navigation, traffic signal control, and weather changes. Figure 5.11a compares the inference time. The microservice-based framework achieves an inference time of less than 0.5 s, meeting real-time interaction requirements and significantly outperforming the traditional approach. Figure 5.11b illustrates the execution accuracy results. The microservice-based method generates accurate traffic codes across all instruction types, while the traditional framework struggles with certain instructions, leading to incorrect or non-executable codes. Figure 5.11c shows the computing cost, where the microservice-based framework mostly achieves lower costs compared to the traditional approach, demonstrating better resource efficiency. Figure 5.11d compares the delay between the traditional approach and the microservice-based framework. The microservice-based approach consistently outperforms the traditional method, achieving significantly lower AVE-DL, MAX-DL, MIN-DL, and JIT-DL. Compared to recent edge computing schemes (without enabling cloud computing and terminal's assistance) in intelligent transportation, our approach leverages a cloud-edge-terminal setup to balance processing loads, enhancing scalability for high-density traffic. This architecture enables efficient large-scale data management, complementing existing purely edge computing based edge-focused solutions.

5.5 Summary of the Chapter

This chapter presents a comprehensive exploration of an innovative cloud-edge-end collaborative framework, SV-FDT, and its extension into an AIGC-driven real-time 4D traffic scene generation system, aiming to revolutionize ITSs by integrating pedestrians, vehicles, and human interactions into a unified loop. The proposed SV-FDT framework incorporates a comprehensive system architecture that emphasizes key design requirements and addresses major challenges for semantic segmentation-based FDT construction. By integrating advanced perception modules, pedestrian-vehicle interaction modeling, and a dynamic digital twin updating mechanism, SV-FDT significantly enhances the fidelity and responsiveness of traffic management systems. Applications such as extreme pedestrian-vehicle flow testbeds and emergency response simulations validate its feasibility. Our experimental results based on CARLA simulations demonstrate that SV-FDT outperforms traditional terminal-server frameworks in terms of mirroring delay, recognition accuracy, and subjective realism.

In addition to the SV-FDT framework, we have introduced an AIGC-driven extension designed for real-time, interactive 4D traffic scene generation. This extension overcomes limitations inherent in conventional text-to-video systems by leveraging text-to-code generation capabilities to orchestrate atomic microservice operations. Integrated within a cloud-edge-terminal collaborative architecture, the system employs multi-modal sensors, speech interfaces, and locally deployed LLMs to support immersive traffic scene generation, enabling real-time responsiveness and user-centric interactions. A microservice-based orchestration mechanism ensures efficient and accurate execution of user instructions in CARLA environments. Evaluation results confirm that this approach delivers superior inference speed, execution accuracy, and system scalability under high concurrency, making it highly suitable for future ITS deployments.

References

1. Akopov, A.S., Beklaryan, L.A.: Traffic improvement in Manhattan road networks with the use of parallel hybrid biobjective genetic algorithm. IEEE Access **12**, 19532–19552 (2024)
2. Akram, B.O., Noordin, N.K., Hashim, F., Rasid, M.A.F., Salman, M.I., Abdulghani, A.M.: Enhancing reliability of time-triggered traffic in joint scheduling and routing optimization within time-sensitive networks. IEEE Access **12**, 78379–78396 (2024)
3. Brooks, T., Peebles, B., Holmes, C., DePue, W., Guo, Y., Jing, L., Schnurr, D., Taylor, J., Luhman, T., Luhman, E., Ng, C., Wang, R., Ramesh, A.: Video generation models as world simulators. (2024). https://openai.com/research/video-generation-models-as-world-simulators
4. Cai, J., Shen, X., Mark, J.W., Alfa, A.S.: Semi-distributed user relaying algorithm for amplify-and-forward wireless relay networks. IEEE Trans. Wirel. Commun. **7**(4), 1348–1357 (2008)
5. Cao, H., Cai, J.: Distributed multiuser computation offloading for cloudlet-based mobile cloud computing: a game-theoretic machine learning approach. IEEE Trans. Veh. Technol. **67**(1), 752–764 (2018)
6. Cao, P., Zhang, C., Huang, J., Wang, Y., Fan, L., Zhao, Y., Zheng, X.: Cross-light contrastive learning of pavement crack segmentation in low-light conditions for autonomous road inspection vehicles. IEEE Trans. Intell. Veh. 1–11 (2024)
7. Chen, J., Yi, C., Du, H., Niyato, D., Kang, J., Cai, J., Shen, X.: A revolution of personalized healthcare: enabling human digital twin with mobile AIGC. IEEE Netw. **38**(6), 234–242 (2024)
8. Chen, J., Yi, C., Okegbile, S.D., Cai, J., Shen, X.S.: Networking architecture and key supporting technologies for human digital twin in personalized healthcare: a comprehensive survey. IEEE Commun. Surv. Tutor. **26**(1), 706–746 (2024)
9. Chen, R., Yi, C.: A game-theoretic online optimization for federated digital twin construction via wireless sensing. In: Proc. IEEE ICC. Montreal, Canada (2025)
10. Chen, R., Yi, C., Zhou, F., Kang, J., Wu, Y., Niyato, D.: Federated digital twin construction via distributed sensing: a game-theoretic online optimization with overlapping coalitions. IEEE Trans. Mobile Comput. 1–17 (2025)
11. Chen, R., Yi, C., Zhu, K., Cai, J., Chen, B.: A DRL-based hierarchical game for physical layer security with dynamic trilateral coalitions. In: Proc. IEEE ICC, pp. 4495–4500 (2023)
12. Chen, R., Yi, C., Zhu, K., Chen, B., Cai, J., Guizani, M.: A three-party hierarchical game for physical layer security aware wireless communications with dynamic trilateral coalitions. IEEE Trans. Wirel. Commun. **23**(5), 4815–4829 (2024)
13. Dong, L., Jiang, F., Wang, M., Peng, Y., Li, X.: Deep progressive reinforcement learning-based flexible resource scheduling framework for IRS and UAV-assisted MEC system. IEEE Trans. Neural Netw. Learn. Syst. **36**(2), 2314–2326 (2025)
14. Guo, Z., Zhang, Y., Lv, J., Liu, Y., Liu, Y.: An online learning collaborative method for traffic forecasting and routing optimization. IEEE Trans. Intell. Transp. Syst. **22**(10), 6634–6645 (2020)

15. Herath, M., Dutta, H., Minerva, R., Crespi, N., Alvi, M., Raza, S.M.: An AI-driven, scalable, and modular digital twin framework for traffic management. In: Proc. IEEE WCNC, pp. 01–06 (2025)
16. Hui, Y., Cheng, N., Su, Z., Huang, Y., Zhao, P., Luan, T.H., Li, C.: Secure and personalized edge computing services in 6G heterogeneous vehicular networks. IEEE Internet Things J. **9**(8), 5920–5931 (2021)
17. Jiang, F., Peng, Y., Dong, L., Wang, K., Yang, K., Pan, C., You, X.: Large AI model-based semantic communications. IEEE Wirel. Commun. **31**(3), 68–75 (2024)
18. Jiang, H., Li, Z., Wei, H., Xiong, X., Ruan, J., Lu, J., Mao, H., Zhao, R.: X-Light: Cross-city traffic signal control using transformer on transformer as meta multi-agent reinforcement learner. arXiv preprint arXiv:2404.12090 (2024)
19. Khan, L.U., Mustafa, E., Shuja, J., Rehman, F., Bilal, K., Han, Z., Hong, C.S.: Federated learning for digital twin-based vehicular networks: architecture and challenges. IEEE Wirel. Commun. **31**(2), 156–162 (2023)
20. Kim, K.S., Kim, D.K., Chae, C.B., Choi, S., Ko, Y.C., Kim, J., Lim, Y.G., Yang, M., Kim, S., Lim, B., et al.: Ultrareliable and low-latency communication techniques for tactile internet services. Proc. IEEE **107**(2), 376–393 (2018)
21. Kuribayashi, M., Uehara, K., Wang, A., Sato, D., Chu, S., Morishima, S.: Memory-Maze: scenario driven benchmark and visual language navigation model for guiding blind people. arXiv preprint arXiv:2405.07060 (2024)
22. Li, G., Cai, J.: An online incentive mechanism for collaborative task offloading in mobile edge computing. IEEE Trans. Wirel. Commun. **19**(1), 624–636 (2020)
23. Li, X., Wei, J., Wang, H., Dong, L., Chen, R., Yi, C., Cai, J., Niyato, D., Shen, X.: Towards intelligent transportation with pedestrians and vehicles in-the-loop: a surveillance video-assisted federated digital twin framework. IEEE Netw. 1-1 (2025)
24. Liang, J., Tan, C., Yan, L., Zhou, J., Yin, G., Yang, K.: Interaction-aware trajectory prediction for safe motion planning in autonomous driving: a transformer-transfer learning approach. IEEE Trans. Intell. Transp. Syst. 1–16 (2025)
25. Liu, D., Chen, R., Yi, C., Zhang, T., Li, X., Li, J., Wang, R., Zhu, K.: A hierarchical game for physical layer security aware cooperative communications with dynamic interchangeable relaying and jamming. IEEE Trans. Veh. Technol. **74**(1), 968–983 (2025)
26. Liu, D., Chen, R., Zhang, T., Yi, C.: A DRL-based hierarchical game for physical layer security aware cooperative communications. In: Proc. IEEE ICCC, pp. 1769–1774 (2024)
27. Lu, B., Huang, X., Wu, Y., Qian, L., Niyato, D., Xu, C.: Cooperative perception aided digital twin model update and migration in mixed vehicular networks. IEEE Trans. Intell. Transp. Syst. **26**(2), 2293–2308 (2025)
28. Luo, Q., Li, C., Luan, T.H., Shi, W.: Minimizing the delay and cost of computation offloading for vehicular edge computing. IEEE Trans. Serv. Comput. **15**(5), 2897–2909 (2021)
29. Man, T., Zhu, H., Chen, R., Yi, C.: Hierarchical DRL-based multi-motor control with torque synchronization in industrial IoT. In: Int. Conf. Wirel. Artif. Intell. Comput. Syst. Appl. (WASA), pp. 11–21. Springer (2025)
30. Manas, K., Zwicklbauer, S., Paschke, A.: TR2MTL: LLM based framework for metric temporal logic formalization of traffic rules. In: Proc. IEEE IV, pp. 1206–1213 (2024)
31. Martinez, V.M., Campelo, D.R., Ribeiro, M.R.: Sustainable intelligent transportation systems via digital twins: a contextualized survey. IEEE Open J. Intell. Transp. Syst. **6**, 363–392 (2025)
32. Masri, S., Ashqar, H.I., Elhenawy, M.: Leveraging large language models (LLMs) for traffic management at urban intersections: the case of mixed traffic scenarios. arXiv preprint arXiv:2408.00948 (2024)
33. Mohajer, A., Hajipour, J., Leung, V.C.: Dynamic offloading in mobile edge computing with traffic-aware network slicing and adaptive TD3 strategy. IEEE Commun. Lett. **29**(1), 95–99 (2024)
34. Nie, J., Jiang, J., Li, Y., Ercisli, S.: Generative AI-enhanced autonomous driving: innovating decision-making and risk assessment in multi-interactive environments. IEEE Trans. Intell. Transp. Syst. 1–16 (2025)

35. Okegbile, S.D., Cai, J., Yi, C., Niyato, D.: Human digital twin for personalized healthcare: vision, architecture and future directions. IEEE Netw. **37**(2), 262–269 (2023)
36. Qian, Z., Chen, R., Yi, C., Zhai, X., Chen, B.: Collision avoidance control for autonomous driving with multiple dynamic obstacles in IoV: a prediction-enhanced apf-based approach. IEEE Internet Things J. (2025)
37. Savkin, A., Ellouze, R., Navab, N., Tombari, F.: Unsupervised traffic scene generation with synthetic 3D scene graphs. In: Proc. IEEE IROS, pp. 1229–1235 (2021)
38. Tang, L., Wen, M., Shan, Z., Li, L., Liu, Q., Chen, Q.: Digital twin-enabled efficient federated learning for collision warning in intelligent driving. IEEE Trans. Intell. Transp. Syst. **25**(3), 2573–2585 (2023)
39. Wang, C., MacDonald, L.E., Jeni, L.A., Lucey, S.: Flow supervision for deformable NeRF. In: Proc. IEEE CVPR, pp. 21128–21137 (2023)
40. Wang, M., Pang, A., Kan, Y., Pun, M.O., Chen, C.S., Huang, B.: LLM-assisted light: leveraging large language model capabilities for human-mimetic traffic signal control in complex urban environments. arXiv preprint arXiv:2403.08337 (2024)
41. Wang, Z., Gupta, R., Han, K., Wang, H., Ganlath, A., Ammar, N., Tiwari, P.: Mobility digital twin: concept, architecture, case study, and future challenges. IEEE Internet Things J. **9**(18), 17452–17467 (2022)
42. Wang, Z., Wang, Y., Jia, F., Zhang, F., Klimenko, N., Wang, L., He, Z., Huang, Z., Liu, Y.: Spatiotemporal fusion transformer for large-scale traffic forecasting. Inform. Fusion **107**, 102293 (2024)
43. Wang, Z., Zheng, O., Li, L., Abdel-Aty, M., Cruz-Neira, C., Islam, Z.: Towards next generation of pedestrian and connected vehicle in-the-loop research: a digital twin co-simulation framework. IEEE Trans. Intell. Veh. **8**(4), 2674–2683 (2023)
44. Wei, C., Wu, G., Barth, M.J., Abdelraouf, A., Gupta, R., Han, K.: KI-GAN: Knowledge-informed generative adversarial networks for enhanced multi-vehicle trajectory forecasting at signalized intersections. In: Proc. IEEE CVPR, pp. 7115–7124 (2024)
45. Wu, J., Chen, R., Yi, C.: A DRL-based deviation-aware federated digital twin construction over wireless edge network. In: Int. Conf. Wirel. Artif. Intell. Comput. Syst. Appl. (WASA), pp. 196–208. Springer (2025)
46. Xu, Q., Chraibi, M., Seyfried, A.: Anticipation in a velocity-based model for pedestrian dynamics. Transp. Res. Part C Emerg. Technol. **133**, 103464 (2021)
47. Yang, Y., Shi, Y., Chen, R., Yi, C., Kang, J.: Online optimization of edge empowered human digital twin deployment and task offloading. In: Proc. IEEE ICCC, pp. 856–861 (2024)
48. Yang, Y., Shi, Y., Yi, C., Cai, J., Kang, J., Niyato, D., Shen, X.: Dynamic human digital twin deployment at the edge for task execution: a two-timescale accuracy-aware online optimization. IEEE Trans. Mob. Comput. **23**(12), 12262–12279 (2024)
49. Yi, C., Cai, J., Su, Z.: A multi-user mobile computation offloading and transmission scheduling mechanism for delay-sensitive applications. IEEE Trans. Mobile Comput. **19**(1), 29–43 (2020)
50. Yi, C., Huang, S., Cai, J.: An incentive mechanism integrating joint power, channel and link management for social-aware d2d content sharing and proactive caching. IEEE Trans. Mobile Comput. **17**(4), 789–802 (2018)
51. Yu, T., Li, Z., Hashash, O., Sakaguchi, K., Saad, W., Debbah, M.: Internet of federated digital twins: connecting twins beyond borders for society 5.0. IEEE Internet Things Mag. **7**(5), 64–71 (2024)
52. Yu, X., Yang, Y., Chen, R., Yi, C.: Online optimization of edge vehicle digital twin migration with adaptive mobility prediction. In: Proc. IEEE VTC. Chengdu, China (2025)
53. Zhang, J., Wei, Z., Liu, B., Wang, X., Yu, Y., Zhang, R.: Cloud-edge-terminal collaborative AIGC for autonomous driving. IEEE Wirel. Commun. **31**(4), 40–47 (2024)
54. Zhang, Y., Lou, S., Hang, P., Huang, W., Yang, L., Lv, C.: Interactive prediction and decision-making for autonomous vehicles: online active learning with traffic entropy minimization. IEEE Trans. Intell. Transp. Syst. **25**(11), 17718–17732 (2024)
55. Zhao, Y., Li, K., Cheng, Z., Qiao, P., Zheng, X., Ji, R., Liu, C., Yuan, L., Chen, J.: GraCo: granularity-controllable interactive segmentation. In: Proc. IEEE CVPR, pp. 3501–3510 (2024)

56. Zheng, Y., Li, X., Nagano, K., Liu, S., Hilliges, O., De Mello, S.: A unified approach for text-and image-guided 4D scene generation. In: Proc. IEEE CVPR, pp. 7300–7309 (2024)
57. Zhou, H., Chen, R., Yi, C., Li, J., Cai, J.: A three-party repeated coalition formation game for pls in wireless communications with irss. In: Proc. IEEE WCNC, pp. 1–6 (2024)
58. Zhou, H., Chen, R., Yi, C., Zhang, J., Kang, J., Cai, J., Guizani, M.: A repeated coalition formation game for physical layer security aware wireless communications with third-party intelligent reflecting surfaces. IEEE Trans. Wirel. Commun. 1-1 (2025)
59. Zhu, W., Chen, R., Yi, C., Cai, J.: Edge-assisted video transmission with adaptive key frame selection: a hierarchical DRL approach. In: Proc. IEEE Biennial Symp. Commun., pp. 89–94. IEEE (2023)
60. Zofka, M.R., Ulbrich, S., Karl, D., Fleck, T., Kohlhaas, R., Rönnau, A., Dillmann, R., Zöllner, J.M.: Traffic participants in the loop: a mixed reality-based interaction testbed for the verification and validation of autonomous vehicles. In: Proc. IEEE ITSC, pp. 3583–3590 (2018)

Chapter 6
Conclusions and Future Research Directions

6.1 Conclusions

In this book, we have undertaken a systematic study of DT over wireless networks along three tightly coupled tracks: i) constructing high-fidelity and evolvable DT models with distributed sensing and edge–cloud collaboration; ii) enabling QoE-aware DT service interaction with uncertain evolution; and iii) implementing a DT system in vehicular network through cloud–edge–end systems and real-time 4D scene generation. Together, these contributions form a coherent foundation that spans the principles, algorithms, and prototypes of DT at scale.

- A comprehensive overview of DT over wireless networks has been presented, covering its concept, architecture, and distinctive features compared with conventional simulation models and general DT services. Representative application scenarios were analyzed to highlight practical values, and the enabling technologies for distributed DT construction and service interaction were identified.
- To meet stringent end-to-end requirements [13, 18], we examined design challenges across data, compute, and network planes: DT construction calls for accurate, synchronized multimodal acquisition and scalable processing; DT interaction demands immersive, low-latency, collaborative experiences. A cross-layer toolbox was articulated, ranging from sensing hardware and edge AI to deterministic, task-oriented communications with RIS, semantic transmission, cell-free massive MIMO, and privacy-preserving edge intelligence under zero-trust security [7, 8].
- We developed a federated DT construction framework supported by distributed sensing and edge-cloud collaboration [5, 6]. An online optimization problem was formulated to jointly determine partial-DT assignments, ES-sensor associations, and resource allocations. By transforming the problem into a hierarchical game and developing a DMO-based approach, we achieved stable and efficient long-term solution, with simulation results validating its effectiveness.

C. Yi et al., *Self-Evolving Digital Twin Over Wireless Networks*, Wireless Networks,
https://doi.org/10.1007/978-3-032-08120-9_6

- For DT service interaction, we studied a QoE-aware resource allocation problem in RIS-assisted wireless networks with uncertain evolutions. By jointly optimizing phase shifts, beamforming, rendering resolution, and computing resources, a GAI-aided algorithm named PG-ZFO was developed [2]. This method effectively balanced objective system performance and subjective user experience, outperforming conventional schemes.
- To demonstrate the potential of DT over wireless networks in future intelligent transportation systems [9], a cloud–edge–end collaborative framework, SV-FDT, together with an AIGC-driven extension, has been introduced. By integrating semantic perception, pedestrian–vehicle interaction modeling, and microservice-based orchestration, the framework enables real-time and interactive 4D traffic scene generation. Experiments conducted in CARLA validate the superior fidelity, responsiveness, and scalability of the proposed framework compared with conventional approaches.

6.2 Future Research Directions

Despite these advances, several open challenges remain and warrant further investigation. We summarize the major future research directions as follows:

- *Robust Federated DT Construction Under Non-i.i.d. Data*: Non-i.i.d. distributed sensing data may contain personalized characteristics from individual sensors, leading to significant deviations in partial-DT models [15]. Developing non-overlapping coalitional game formulations [10, 21] for ES–sensor association, where each sensor is exclusively assigned to a single ES, can help ensure model consistency.
- *Redundancy Elimination in Overlapping Feature Data*: For the same partial-DT creation, feature data collected by different sensors may overlap, introducing redundancy and degrading the quality of the global DT. Advanced data preprocessing techniques and sensor-state prediction algorithms should be designed to eliminate redundancy and improve robustness.
- *RIS-Assisted DT Interaction Under Hardware Constraints*: Discrete phase shift control for RIS [20], along with quantization-aware training, should be studied to reflect practical hardware limitations and enable more accurate DT service interaction in wireless networks.
- *Advanced Channel Modeling and Estimation*: To better represent real-world wireless environments, imperfect CSI and sophisticated channel fading models need to be incorporated [1, 22]. This requires reformulating MDPs and combining them with deep learning–based channel estimation techniques.
- *Event-Driven User Interaction Modeling*: User interaction requests in DT systems are often event-driven and mobile [19], with unpredictable variations in participant numbers. Designing scalable mechanisms to handle such variations is an important direction.

- *Robustness Under Uncertainty*: External environmental factors may violate the Markov property [4, 11], requiring more resilient modeling tools. Partially observable MDPs and causal graph–based models represent promising approaches to enhance robustness in uncertain environments.

Building upon the theoretical research on DT construction and service interaction, we further explore their implementation in practical applications. In particular, we focus on DT deployments in ITS, highlighting challenges such as multimodal data precision, real-time responsiveness, scalability, and user-centric interaction, as detailed below.

- *High-Precision DT Extraction via Multimodal Fusion*: Future work should enhance the precision of digital twins by integrating and fusing multimodal data (e.g., images, audio, text, video). Real-time acquisition and distributed storage of such data raise issues of latency, privacy, and security. Emerging techniques such as knowledge graphs, deep learning-based feature alignment, and secure data sharing protocols will be explored to build more accurate and secure DT models.
- *Optimizing Data Fusion and Processing*: In the context of AIGC-driven scene generation [3], research can also focus on advanced techniques for integrating and processing heterogeneous ITS data using deep learning, edge computing, and blockchain-like distributed ledger technologies. These can ensure real-time responsiveness while maintaining system integrity and data privacy.
- *Enhancing Interactive Experiences*: To improve user engagement in virtual transportation platforms, future work can explore human-in-the-loop interfaces such as virtual reality (VR), augmented reality (AR), and advanced NLP-based voice interfaces. These technologies will enable more intuitive, immersive control and observation of dynamic 4D traffic scenes.
- *Extending DT to Future ITS with Hyper-Spatial Mobility*: With the evolution of AI and transportation systems, ITS will evolve into hyper-spatial ecosystems encompassing road, railway, air, and waterway mobility. Future research will focus on intelligent control, autonomous decision-making, and resource-aware modeling using algorithms such as TD3 and deep reinforcement learning [12]. These tools will support unified route planning, state monitoring, and predictive fault detection in multi-modal networks.
- *Balancing Real-time Performance with Scalability*: As system complexity grows, real-time scene generation and processing must be balanced with system scalability. Optimizing microservice architecture [14], container orchestration, and edge-cloud resource scheduling [17] will be key to handling high-concurrency ITS deployments.
- *Optimal Traffic Signal Control in Self-Evolving DT Systems*: Inaccurate DT representations can degrade traffic signal control performance. Future research will explore robust filtering techniques, continual learning, and multi-agent reinforcement learning strategies to enable accurate, self-evolving DT coordination [16]. Local agents will adaptively optimize signal timing based on real-time pedestrian-vehicle interactions, while privacy-preserving model updates will be aggregated

at the cloud level for global decision-making. Semantic-level interaction modeling and probabilistic behavior inference mechanisms will be used to further refine agent behavior in uncertain traffic environments.

References

1. Cai, J., Shen, X., Mark, J.W., Alfa, A.S.: Semi-distributed user relaying algorithm for amplify-and-forward wireless relay networks. IEEE Trans. Wirel. Commun. **7**(4), 1348–1357 (2008)
2. Chen, J., Shi, Y., Yi, C., Du, H., Kang, J., Niyato, D.: Generative AI-driven human digital twin in IoT-healthcare: a comprehensive survey. IEEE Internet Things J. **11**(21), 34749–34773 (2024)
3. Chen, J., Yi, C., Du, H., Niyato, D., Kang, J., Cai, J., Shen, X.: A revolution of personalized healthcare: enabling human digital twin with mobile AIGC. IEEE Netw. **38**(6), 234–242 (2024)
4. Chen, J., Yi, C., Wang, R., Zhu, K., Cai, J.: Learning aided joint sensor activation and mobile charging vehicle scheduling for energy-efficient WRSN-based industrial IoT. IEEE Trans. Veh. Technol. **72**(4), 5064–5078 (2023)
5. Chen, R., Yi, C.: A game-theoretic online optimization for federated digital twin construction via wireless sensing. In: Proceedings of IEEE ICC, Montreal, Canada (2025)
6. Chen, R., Yi, C., Zhou, F., Kang, J., Wu, Y., Niyato, D.: Federated digital twin construction via distributed sensing: a game-theoretic online optimization with overlapping coalitions. IEEE Trans. Mobile Comput. 1–17 (2025)
7. Chen, R., Yi, C., Zhu, K., Cai, J., Chen, B.: A DRL-based hierarchical game for physical layer security with dynamic trilateral coalitions. In: Proceedings of IEEE ICC, pp. 4495–4500 (2023)
8. Chen, R., Yi, C., Zhu, K., Chen, B., Cai, J., Guizani, M.: A three-party hierarchical game for physical layer security aware wireless communications with dynamic trilateral coalitions. IEEE Trans. Wirel. Commun. **23**(5), 4815–4829 (2024)
9. Li, X., Wei, J., Wang, H., Dong, L., Chen, R., Yi, C., Cai, J., Niyato, D., Shen, X.: Towards intelligent transportation with pedestrians and vehicles in-the-loop: a surveillance video-assisted federated digital twin framework. IEEE Netw. 1–1 (2025)
10. Liu, D., Chen, R., Yi, C., Zhang, T., Li, X., Li, J., Wang, R., Zhu, K.: A hierarchical game for physical layer security aware cooperative communications with dynamic interchangeable relaying and jamming. IEEE Trans. Veh. Technol. **74**(1), 968–983 (2025)
11. Liu, D., Chen, R., Zhang, T., Yi, C.: A DRL-based hierarchical game for physical layer security aware cooperative communications. In: Proceedings of IEEE ICCC, pp. 1769–1774 (2024)
12. Man, T., Zhu, H., Chen, R., Yi, C.: Hierarchical DRL-based multi-motor control with torque synchronization in industrial IoT. In: International Conference on Wireless Artificial Intelligent Computing Systems and Applications (WASA), pp. 11–21. Springer (2025)
13. Okegbile, S.D., Cai, J., Yi, C., Niyato, D.: Human digital twin for personalized healthcare: vision, architecture and future directions. IEEE Netw. **37**(2), 262–269 (2023)
14. Shi, Y., Yang, Y., Yi, C., Chen, B., Cai, J.: Towards online reliability-enhanced microservice deployment with layer sharing in edge computing. IEEE Internet Things J. **11**(13), 23370–23383 (2024)
15. Wu, J., Chen, R., Yi, C.: A DRL-based deviation-aware federated digital twin construction over wireless edge network. In: International Conference on Wireless Artificial Intelligent Computing Systems and Applications (WASA), pp. 196–208. Springer (2025)
16. Yang, Y., Shi, Y., Yi, C., Cai, J., Kang, J., Niyato, D., Shen, X.: Dynamic human digital twin deployment at the edge for task execution: a two-timescale accuracy-aware online optimization. IEEE Trans. Mobile Comput. **23**(12), 12262–12279 (2024)
17. Yi, C., Cai, J., Su, Z.: A multi-user mobile computation offloading and transmission scheduling mechanism for delay-sensitive applications. IEEE Trans. Mob. Comput. **19**(1), 29–43 (2020)

18. Yi, C., Huang, S., Cai, J.: An incentive mechanism integrating joint power, channel and link management for social-aware D2D content sharing and proactive caching. IEEE Trans. Mob. Comput. **17**(4), 789–802 (2018)
19. Yu, X., Yang, Y., Chen, R., Yi, C.: Online optimization of edge vehicle digital twin migration with adaptive mobility prediction. In: Proceedings of IEEE VTC, Chengdu, China (2025)
20. Zhou, H., Chen, R., Yi, C., Li, J., Cai, J.: A three-party repeated coalition formation game for PLS in wireless communications with IRSS. In: Proceedings of IEEE WCNC, pp. 1–6 (2024)
21. Zhou, H., Chen, R., Yi, C., Zhang, J., Kang, J., Cai, J., Guizani, M.: A repeated coalition formation game for physical layer security aware wireless communications with third-party intelligent reflecting surfaces. In: IEEE Transactions of Wireless Communication, pp. 1–1 (2025)
22. Zhu, W., Chen, R., Yi, C., Cai, J.: Edge-assisted video transmission with adaptive key frame selection: a hierarchical DRL approach. In: Proceedings of IEEE Biennial Symposium on Communications, pp. 89–94. IEEE (2023)

Index

C. Yi et al., *Self-Evolving Digital Twin Over Wireless Networks*, Wireless Networks,
https://doi.org/10.1007/978-3-032-08120-9